THE
TYNE DEE
COASTAL
No 4498

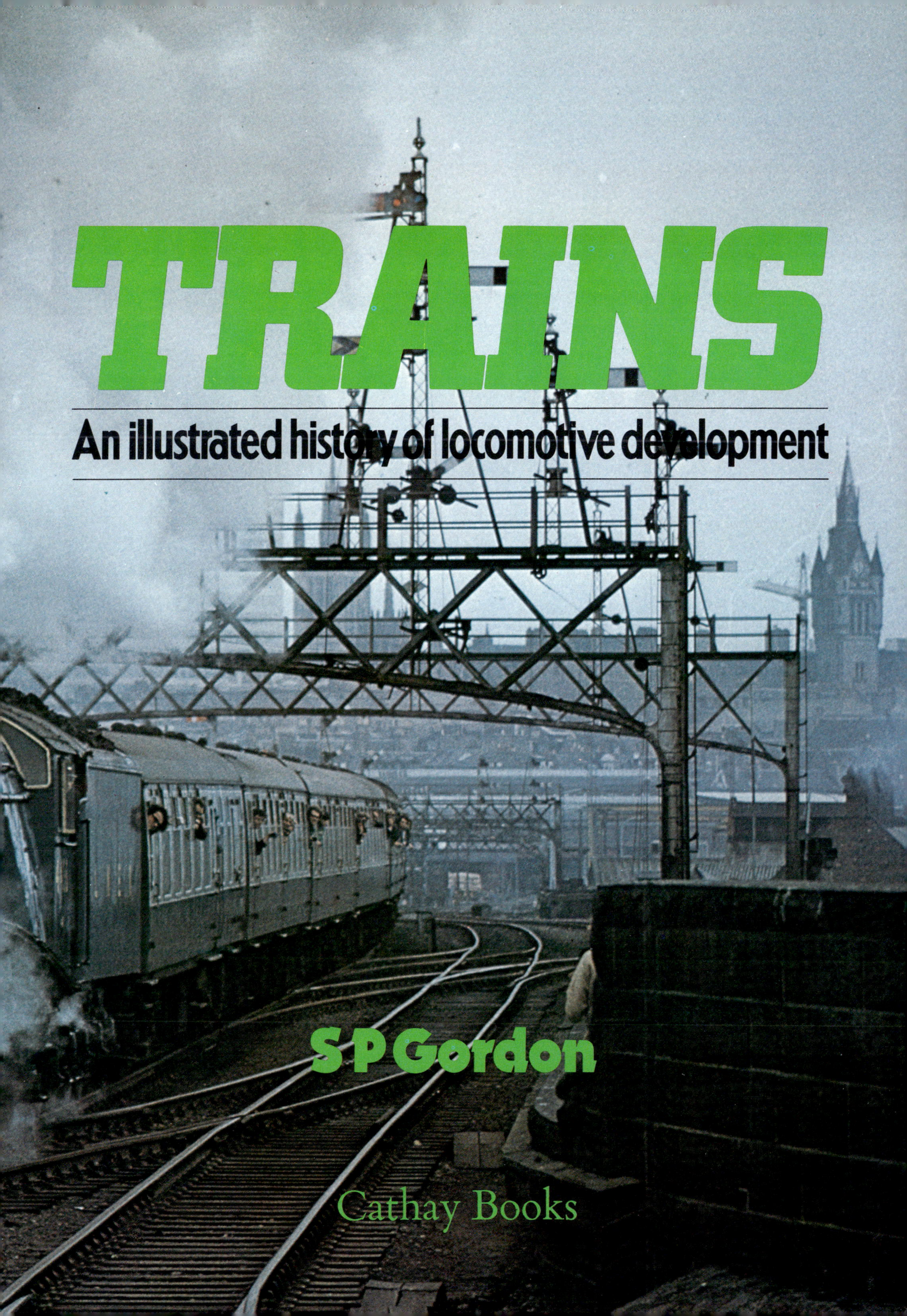
TRAINS
An illustrated history of locomotive development
S P Gordon
Cathay Books

TRAINS

CONTENTS

First published 1976 by Cathay Books Ltd, 59 Grosvenor Street, London W1

ISBN 0 904644 19 7
Produced by Mandarin Publishers Limited 22A Westlands Road, Quarry Bay, Hong Kong
Printed in Hong Kong

STEAM ESTABLISHED IN BRITAIN

One of the first steam engines – the 'Puffing Billy' – built by William Hedley and Christopher Blackett in 1813 for use at the Wylam Colliery, Newcastle.

Railways predate the advent of locomotives by more than two centuries for they have been in existence, in some crude form, for over four hundred years. In the middle of the sixteenth century, a German book was published showing a miner pulling a wooden truck along a system of wooden runways or 'rails'. Similar rails appear to have been constructed at a mine in Nottingham in 1597, and in 1680 Roger North wrote of 'the laying of rails of timber exactly straight and parallel' that he had seen on the roads leading to and from the collieries around Newcastle-on-Tyne. These early wooden railways, and others made of stone, were designed to provide a smooth path for the wheels of wagons pulled by horses, which plodded the rougher road surface between them.

The timber tracks were quick to wear and rot away so the refinement of fixing metal plates was introduced. Soon another refinement was added – flanges to keep the wheels of the wagons on a straight path. In 1767 a famous ironmaster – Abraham Darby – built the first iron rails which were very gradually adopted in places where metal was plentiful. For many years to come, however, the metal-flanged, wooden rails were to be used and the first public railway in the world, from Croydon to Wandsworth opened in 1804, was of such a system.

Although the rumblings – or more accurately the deafening roars – of self-powered wagons, or locomotives, could be heard in some parts of the country by then, the Croydon–Wandsworth railway still used horse-drawn wagons. It was, however, in the February of the same year that the first railway locomotive, the founder of a family that before long would traverse the continents of the world, jerked its way along the wood-based tramways. The locomotive was the brainchild of a Cornish inventor Richard Trevithick, sometimes called the Father of the Locomotive, but whose name often occupies a less prominent place than the later masters of steam locomotion.

The use of steam to power stationary machines was not new. There are accounts, admittedly not indisputably authentic, of a boat with paddle wheels powered by a steam engine that was exhibited on the harbour of Barcelona in 1543. A hundred years later an Italian engineer produced a steam-powered 'windmill', and in the early eighteenth century a Devonshire blacksmith perfected an 'atmosphere engine' to pump water out of mines. James Watt brought the steam engine nearer perfection later the same century, thus bringing it also into greater repute.

Richard Trevithick, however, was the first man to develop a high-pressure, stationary engine as opposed to the more cumbersome, low-pressure engine of high coal consumption. Instead of driving the piston by the weight of air above it, he used pressurized steam alone. It was this invention that he adapted for his locomotive, making it move along by power transmitted to the axles through spun wheels. To describe it crudely, it was a large boiler, firebox and cylinder with a small smokestack set on four wheels. A horizontal cylinder inside the boiler was linked to a projective slide base which drove the crankshaft attached to a huge flywheel. Exhaust steam was discharged into the chimney to draw up the fire. It took to the tracks at Penydaren Ironworks, near Merthyr Tydfil in South Wales, where it required four hours to cover 10 miles of track, hauling 10 tons of iron ore. It was in addition the subject of a 500-guinea bet, for Anthony Hill, the owner of a nearby ironworks, had made such a wager against the master of Penydaren, claiming the engine could not undertake this task. Amid much hissing and blasting of steam, rattling and screeching, breaking of the rail plates and the cutting down of overhead branches that impeded progress, Anthony Hill lost his 500 guineas.

Sadly, Richard Trevithick never really received the acclaim he deserved. In some ways his steam locomotive was almost ahead of its time, in that track sufficiently strong to bear the weight it demanded had not been developed. 'Uncle Dick's Puffer', as it was christened, was soon taken off its wheels and ended its life ignominiously driving a hammer. Trevithick continued to develop his steam locomotive and in 1808 he even took one, the 'Catch-me-who-can', to London, calling it the Racing Steam Horse and setting it up as a kind of show line. On a circular track laid near Euston Square, he challenged horses to a race. The 8-ton engine ran at 12 mph, but a derailment led to the show being dismantled, and to Trevithick becoming discouraged from further experimentive development along these lines.

Public opinion of the steam locomotive was both sceptical and hostile at this time, with people even regarding it as 'a device of the devil'. We are told that "pamphlets were written and newspapers were hired to revile the railway. It would prevent cows from grazing and hens from laying. The poisoned air it emitted would kill birds, pheasants and foxes. Houses adjoining the lines would be burnt up by the fire thrown from the engine chimneys, while the air around would be polluted by clouds of smoke. There would no longer be any use for horses, which would be likely to become extinct, thus making oats and hay unsaleable commodities.

Richard Trevithick's Railroad Euston Square 1809

Left: A pioneer of steam traction, Richard Trevithick, set up a circular railway in Euston in the early 1800s to demonstrate a steam engine at work. The amazed public are reputed to have paid a shilling a head for admittance to the show – a fee which entitled them to a ride in the carriage if they were brave enough!

Below: The most famous steam engine of all time – Stephenson's 'Rocket'. The subject of many re-builds, it was originally painted bright yellow with a white smokestack and had sloping cylinders which were said to resemble the legs of a grasshopper.

Travelling by road would be made dangerous – and (worse still!) country inns would be ruined. Boilers would burst, blowing passengers to atoms!" A prophesy of doom indeed, but its protagonists took comfort in their knowledge that 'the weight of the locomotive would completely prevent its moving and the railways, if made, could never be worked efficiently by steam power'!

However, steam locomotion was to develop and seven years after Trevithick's engine ran at Penydaren, John Blenkinsop, an employee of a colliery in Middleton, near Leeds, was working to disprove one of the current theories among inventors. This was that locomotives working by adhesion only would fail – on the basis that smooth wheels would slip on smooth rails. In addition there was the problem of the insufficiently strong rails to be overcome. Blenkinsop went part way to solving both

these problems by developing a wooden wagonway with iron-flanged, 'racked' rails, along which he ran a locomotive constructed with the help of Matthew Murray. It was similar in design to Trevithick's, except it had a cogged driving wheel that engaged with pegs fitted to the line. Blenkinsop and Murray produced this 'rack' engine, called the 'Prince Regent', in 1812 and it was used on the Middleton colliery tramway in Leeds. Able to pull eight wagons carrying loads of over 3 tons at 3 mph, or travel free at 10 mph, 'Prince Regent' was soon joined by two other locomotives built by Murray. It is interesting to note that the principle of Blenkinsop's 'racked rail' is still in use on some steep-grade mountain railways.

Other people were still working on the straight adhesion locomotive and in 1813, William Hedley and Christopher Blackett built the original famous 'Puffing Billy' and its sister engine 'Wylam Dilly'. 'Puffing Billy' resembled a grasshopper with four smooth wheels and vertical cylinders and it was installed at Wylam colliery in Newcastle. Here it did the work of ten horses by hauling eight loaded coal wagons at 10 mph. Even though the wooden rails at Wylam had been replaced by cast-iron plates to allow for steam haulage the track was still inadequate for the locomotives. Derailments and breaking of the track were numerous. After a while Hedley mounted 'Puffing Billy' on eight wheels instead of four in an effort to overcome this problem. It seemed it went far towards doing this, for 'Puffing Billy' was said to work satisfactorily in this form for many years.

William Hedley therefore proved that the weight

of an engine would be sufficient to make the whole adhere to rails, and that friction alone would ensure progression without slipping. His trains had been running a year before one of the great names of steam locomotion, George Stephenson, took up the cause.

Stephenson is another to be bestowed with the title of Parent of the Locomotive Engine. Although his success in developing it, making it more efficient and generally putting it 'on the map' is undeniable, his role in fact was to put the seal on the pioneering efforts of those who went before him as well as on his own. At the time when the 'Prince Regent' was running along the Middleton Colliery Railway, Stephenson was engine wright at Killingworth High Pit, owned by Lord Ravensworth. In ten months he made his famous engine, first called 'My Lord' and

Left above: 'Locomotion' – the Stephenson engine whose inaugural run marked the opening of the first public railway in the world to use steam engines. This was the Stockton and Darlington Railway, opened in 1825. 'Locomotion' subsequently took part in the Railway Jubilee of 1875 and the Railway Centenary Celebrations in 1925. She was shown at the British Empire exhibition in 1924 and a fully-working replica was built for the 150-year celebrations in 1975.

Left below: A late example of Stephenson's 'Patentee' design. Stephenson first developed this 2-2-2 engine in the early 1830s and followed it with improved versions in subsequent years. It was to form the basic design of the British single-wheeler passenger engine for the next half century.

Below: Timothy Hackworth is often credited with producing the Stockton and Darlington Railway's first successful engine – the 'Royal George'. The basic design of this engine, of which the most notable characteristic was the presence of a tender at each end, was subjected to many variations and the 'Derwent' (seen below) was one of the concluding forms.

later renamed 'Blücher'. It was similar to the designs of Hedley and Blackett, but it was unique in having flanged wheels. Like other locomotives of the day it was short of steam, possibly because its waste steam was ejected directly into the air. Stephenson soon modified this by directing the waste steam into the chimney to improve the blast and increase the intensity of the fire.

Still the idea of steam locomotion had not really caught on and for the next decade, Stephenson was almost the only person continuing to build steam locomotives. He was convinced of the future of steam in this field and it was his determination as much as his engineering skill that drove him on against general opposition to the 'iron horse'. He developed and patented an improved type of cast-iron rail and another Northumbrian, John Birkinshaw, also developed a new process for making wrought-iron rails. Both inventions were more able to sustain the weight of the heavy engines. By the time Stephenson left the Killingworth mine, he had built sixteen locomotives and had been responsible for laying eight new rail tracks.

At about the same time a rich landowner, Edward Pease, was trying very hard to get Parliament to consent to the building of a public railway from Stockton to Darlington to assist in the mining of the coal seams in County Durham. In 1821 the authorization was granted, with the Act stating that the company was to convey its traffic along the line by 'men, horse or otherwise'. It seems Mr Pease's original intention had been to use wagons and horses, but George Stephenson went to see him and suggested that the 'otherwise' of the Act should be interpreted as steam locomotion. Pease, impressed by Stephenson's earnestness and also the accounts he had heard of him, went to Killingworth to see his engines at work. It was sufficient to convince him and he persuaded his associates that a railroad as opposed to a tramline should be constructed, with

Stephenson appointed as engineer of the project.

George Stephenson was now working very closely with his son Robert who, although not twenty years old, had shown signs of possessing brilliant engineering talent. Together they carried out surveys for the line and in 1823 George persuaded Pease to join them in establishing a locomotive factory. The firm was called Robert Stephenson & Co., and its aim was to carry out experimental work on locomotives, as well as produce them for the Stockton to Darlington run. Although the offer of an exciting job and ill health combined to make Robert leave for a three-year contract in the United States in June 1824, work on the first engine, 'Locomotion', had then neared completion.

George Stephenson decided that when the train of wagons made its first trip, he would also attach a passenger carriage. Then people would see that railroad traction had other uses than just conveying coal or merchandise. To this end an elegant coach named 'Experiment' was designed for the directors' use.

In September 1825, the Stephenson engine 'Locomotion' opened the Stockton and Darlington Railway by pulling twelve wagons of coal, 'Experiment' and twenty-one other trucks, fitted with seats. In all, about 600 passengers rode the 27 miles of track at a speed of 8 mph, making the Stockton and Darlington Railway the first 'public' railway on which locomotive engines were used.

Over the next few years, the Stockton to Darlington line served somewhat as a testing ground for new locomotives. Timothy Hackworth worked to improve the Stephenson engines and produced 'Royal George' two years after the line was opened.

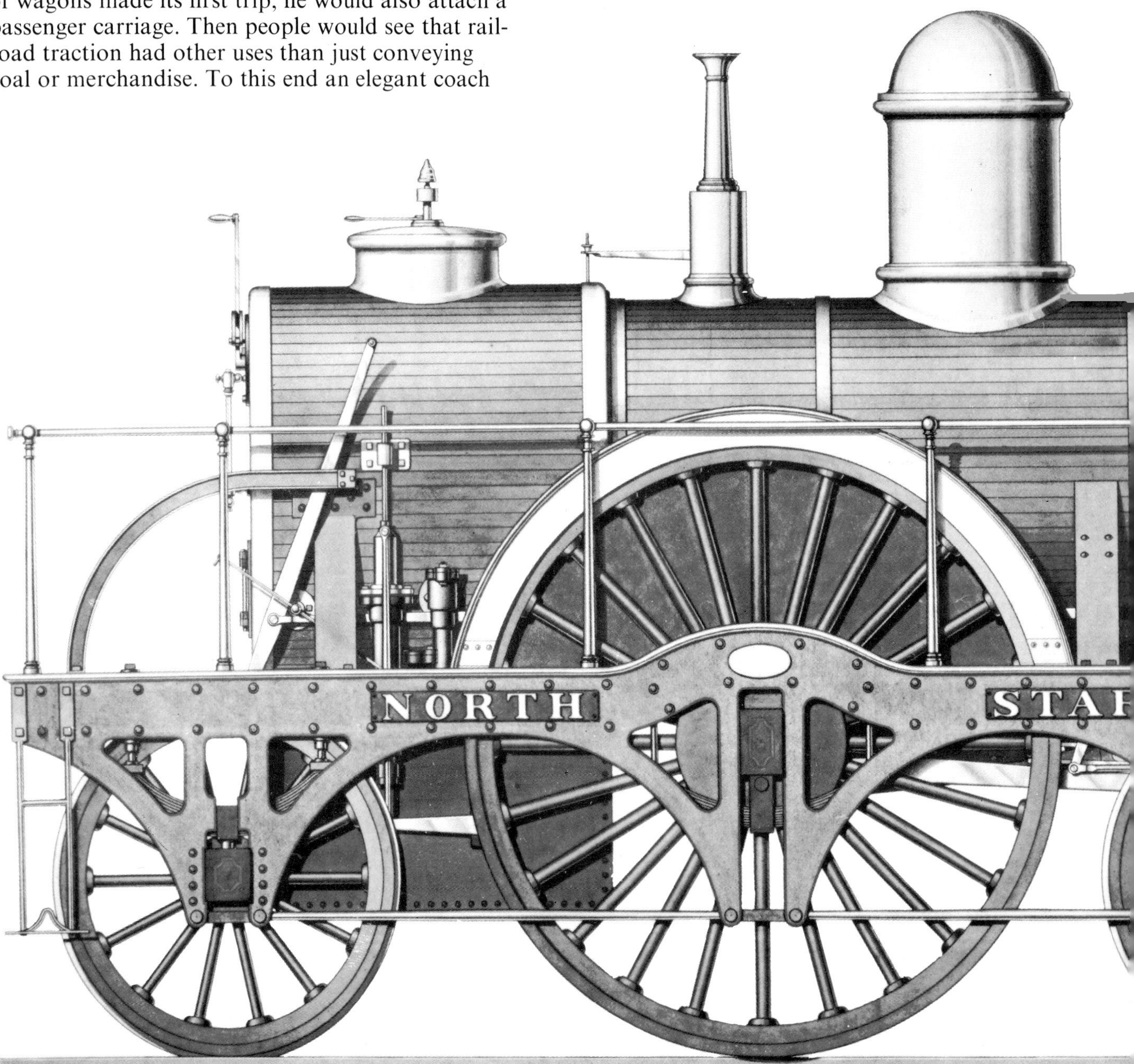

The 'North Star' was built by Stephenson for the broad gauge track of Isambard Brunel's Great Western Railway which opened in 1838.

It had six wheels and a system of compensated levers designed to give some degree of springing.

Perhaps as a result of the Stockton to Darlington success, a project that had lain dormant for a while was revived. This was the Manchester to Liverpool line. The businessmen campaigning for this venture were in favour of steam but were opposed by both canal and landowners. A parliamentary inquiry was held and in spite of the almost panic-stricken opposition, a bill was passed giving the go-ahead in May 1826. George Stephenson was appointed chief engineer to the line.

The engineering difficulties he had to overcome in actually laying the line were enormous. At one place it had to cross a virtual quagmire which repeatedly frustrated all efforts to make it firm. Finally it was beaten. Other engineering triumphs of the line were a 2-mile cutting that had to be dug and blasted at times to a depth of 100 feet; a 70-foot-high nine-arch viaduct, and a 2,250-yard tunnel.

While George Stephenson engineered the line, his son Robert, back from America, was at work again designing engines. In 1828 he produced the 'Lancashire Witch' which ran on the Bolton and Leigh railway. Able to haul a 50-ton load at 8 mph, up a 1 in 440 gradient, it had four coupled wheels driven by two inclined cylinders. Both of its axles were sprung. At the same time the work of a Frenchman, Marc Seguin, was having a considerable effect on engine design. In 1827 he invented a 'multi-tubular' boiler, which was far in advance of the simple boiler of the time and did much to enhance the possibility of speedy long-distance travel.

As work on the Manchester to Liverpool line was nearing completion in 1829 the mode of travel to be used was still not completely decided. Many favoured a system of stationary engines spaced out along the track which would pull the wagons along by cable. The Stephensons, of course, stuck adamantly to their conviction that locomotives should be used and their complete faith made the directors decide to hold a competition with a £500 prize for the best locomotive. Specifications were exact: among other factors, engines were not to exceed 6 tons in weight, they must consume their own smoke and be able to pull three times their own weight at a speed of 10 mph. The £500 prize coupled with the assured purchase of the winning engine and a contract for its designer to supply the locomotives for the line, was to be competed for over a $1\frac{1}{2}$-mile section of the new track, 9 miles from Liverpool and known as the Rainhill level.

Four engines complied with the specified requirements. 'Sanspareil' by Timothy Hackworth, which was somewhat similar in appearance to Trevithick's earlier 'Catch-me-who-can' but contained an important new principle – that of coupled driving wheels; Timothy Burstall's 'Perseverance', which was a carriage with a boiler and chimney in the middle; John Braithwaite and John Ericsson's 'Novelty' resembling an old horse-drawn fire engine with a 'tea-urn' at one end; and the Stephensons' famous 'Rocket', painted yellow and given a 'grasshopper' appearance by its sloping cylinders. It incorporated the multi-tubular boiler, consisting of a series of copper tubes which conveyed the hot gases from the boiler firebox to a smokebox.

Each engine had to make twenty trips averaging not less than 10 mph at the October trials. 'Sans-pareil' apparently 'disappeared in a cloud of steam' on its eighth trip; 'Perseverance' was damaged on its way to the trials and even when repaired was withdrawn; 'Novelty' made several successful runs before being shaken by an explosion caused by her bellows bursting. It was all up to the Stephensons' 'Rocket', which proved itself equal to the occasion by recording a trouble-free average speed of 14 mph over 60 miles (with a top speed on one run of 29 mph). The prize was secured for 'Rocket'.

The Liverpool to Manchester line was opened on 15 September, 1830 with eight of the Stephensons' engines pulling twenty-three carriages. Among the engines were the 'Rocket', the 'Northumbrian' driven by George Stephenson and pulling the Duke of Wellington's (the then Prime Minister) own carriage, and the 'Phoenix' driven by Robert Stephenson. Unfortunately after a halt during the course of the journey, a Member of Parliament, Mr Huskisson, was knocked onto the line by the door of the Duke's carriage and was unable to move in time to escape the approaching 'Rocket'. His leg was crushed, and although George Stephenson conveyed him the 15 miles to Liverpool at a record 36 mph he died shortly after. The line that truly marked the beginning of the railway age had claimed its first victim.

During the construction of the Liverpool to Manchester line, work was being carried out on other comparatively short local runs, such as the Canterbury to Whitstable line. Its first locomotive was a sister design to the 'Rocket' called the 'Invicta'.

The next ten to twenty years was the era of the 'Railway Mania', with people sinking large sums of money into the construction of railways. Although the depressed state of British industry in the decade of 1830–40 meant actual construction work was at times delayed, it saw the formation of many 'railway' companies and the authorization from Parliament for hundreds of miles of track. Notable among these were Isambard Brunel's famous Great Western Railway and one of Stephenson's new ventures from London to Birmingham which was to link with the Liverpool and Manchester line to form the Grand Junction Railway. There were plans, too, for links between Manchester and Leeds, and London and Southampton as well as tracks to service the Midlands, Eastern Counties and the South East.

Right above: Aestheticism in locomotive design was first portrayed in the graceful 'Jenny Lind' class, shown here in a scale drawing.

Right below: Officially named the 'Liverpool' where its designer, Edward Bury had his engineering firm, the high-dome of this engine earned it its commonly used nickname 'Copperknob'. Bury produced it at a time when four-wheeled engines were finding less favour than the new six wheelers.

The construction of Brunel's Great Western Railway and the 'Battle of the Gauges' that resulted is an interesting story in railway history. Of French ancestry, Brunel was appointed chief engineer to the line in 1838 and straight away broke with tradition by laying the rails 7 feet apart, rather than the 4 feet $8\frac{1}{2}$ inches so far favoured by Stephenson. Brunel's railway was, however, a masterpiece and he overcame all the difficulties he encountered by constructing magnificent bridges, tunnels longer than any before built and viaducts or 'spider bridges'. He chose to lay 7-foot-wide rails because he felt them to be safer, more comfortable and able to convey locomotives at higher speed.

While the railways were more or less distinct from one another the difference of gauge did not matter too much, but inevitably links were formed before long. This meant that when the two gauges met, people and goods had to change trains. The result was confusion, delays, and considerable expense in handling goods. In some places along the Great Western Railway route a third rail was laid so that narrow-gauge vehicles could be accommodated as well, but this too was costly.

A government commission was appointed to determine which width was the best. Brunel suggested a speed test and proved indisputably, with a maximum speed of 60 mph, that his track could support trains more safely at faster speeds. During the competition one of Stephenson's locomotives overturned on the narrower gauge at 45 mph. However, this was not enough to convince the government commission, particularly after its members witnessed the confusion at stations when change-overs were being effected. In 1848 the Gauge Act proclaimed the narrower gauge as the approved standard. Doubtless the fact that by then there were nearly 2,000 miles of 'standard' gauge laid against less than 300 of the broad gauge contributed to their decision – but in any event they banned the extension of the broad gauge anywhere except on the Great Western Railway. A little over ten years later, the GWR began to convert its tracks, but the broad track lingered over the next

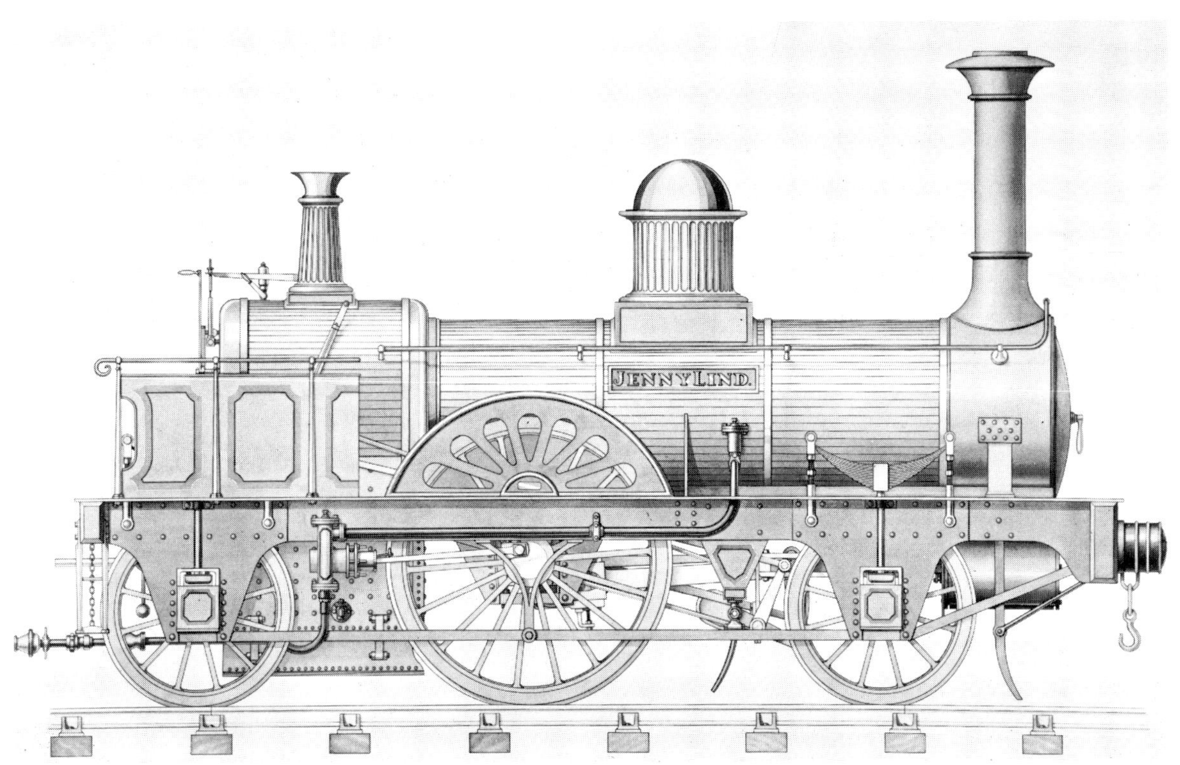
JENNY LIND

forty years, ending with a final frantic conversion period in May 1892.

In 1838 a section of Stephenson's track from London to Birmingham was opened and the Grand Junction Railway also came into service. In addition the London to Southampton line was in operation, but the big boom came in the 1840s when there were vast increases in trade throughout the country. From 1844 to 1846 Parliament gave permission for 438 new lines, which accounted for 8,500 miles of track. More and more companies sprang up and lines to remote little towns and villages were planned.

A name notoriously linked with the Railway Mania of the age was that of George Hudson, known for a long time as 'The Railway King'. He was responsible for most of the activity in the North and Midlands, where together with Stephenson he pioneered many new lines. Unfortunately Hudson was unable to live up to all the extravagances and financial commitments of his ventures and like thousands of others carried away by the Railway Mania, he became a ruined man.

The next years saw the continual building of railway lines and the amalgamation of many companies to form bigger conglomerates. This was attributable in part to the desire to end the vagaries of price competition. But the fall of Hudson in the mid-1840s brought an obvious element of caution into railway affairs and at the end of the 1840s and beginning of the '50s, companies who had been committing themselves to massive expansion a few years earlier began to back-pedal. A brief promotion boom in 1852 and 1853 led to only a modest 1,400 miles of track authorized for building in 1853 and 1854.

During the 1830s and the years of the Railway Mania, the engine designers were far from idle. The 'Northumbrian' built by Robert Stephenson for the Liverpool to Manchester line, although similar to his 'Rocket', had twice its heating surface as well as horizontally placed cylinders to promote greater steadiness. Although both possessed the essential components of all steam locomotives – namely a firebox and a boiler to turn water to steam, cylinders to drive the pistons, and wheels and a chassis – they were virtually out of date by the end of the 1830s when Stephenson produced his 'Planet' locomotive. Here truly was the ancestor of all locomotives that were to follow. Its cylinders were placed inside the smokebox and worked directly on a double-cranked driving axle connected to the rear wheels. The boiler was supported on a triple frame, so that in the event of an axle breaking, the engine would not collapse. For hauling passenger trains it had two small front leading wheels, two larger driving wheels connected to the pistons, and no trailing wheels; this arrangement is referred to as a 2–2–0. A modified version

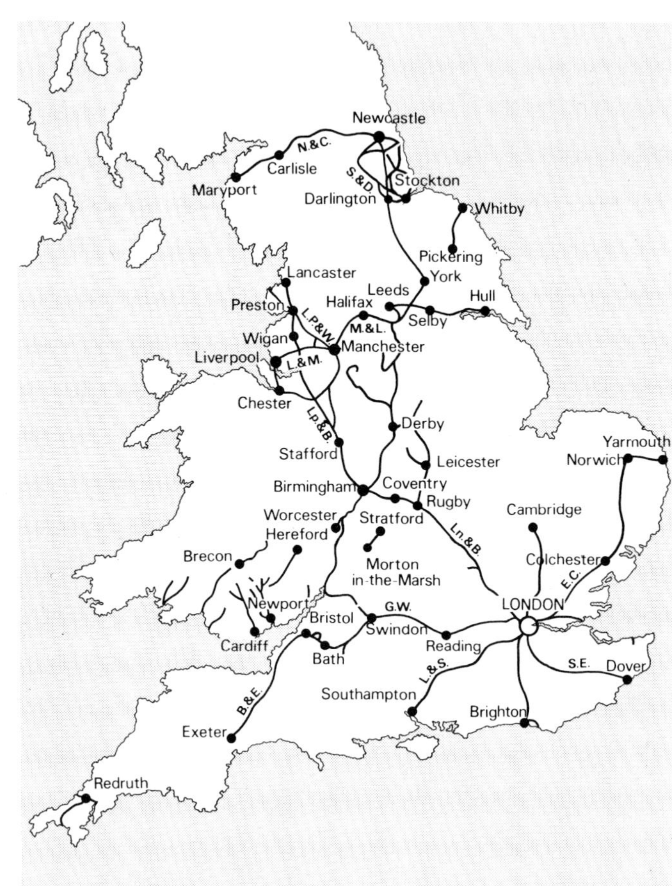

N. and C. – Newcastle and Carlisle.
S. and D. – Stockton and Darlington.
L.P. and W. – Lancaster, Preston and Wigan.
L. and M. – Liverpool and Manchester.
M. and L. – Manchester and Leeds.
Lp. and B. – Liverpool and Birmingham.
Ln. and B. – London and Birmingham.
G.W. – Great Western.
L. and S. – London and Southampton.
E.C. – Eastern Counties.
S.E. – South Eastern.
B. and E. – Bristol and Exeter.

was designed to haul goods and had all four wheels coupled to the drive, which made it an 0–4–0.

The following year Stephenson developed another engine, his 'Patentee', in an effort to overcome the oscillation experienced in the four-coupled engines travelling at even modest speeds. In addition, heavier engines were needed to handle the heavier loads that accompanied the growing popularity of rail travel. As a 2–2–2 'Patentee' had large central driving wheels, with small carrying wheels at either end, it was soon much in demand for pioneer railways in continental Europe. Stephenson continued to improve it, and in 1835, the locomotives of this basic design had driving wheels of 5 feet or more in diameter, a boiler with a heating surface area of 400 square feet and were able to pull a 50-ton load at an average of 20 to 25 mph.

Throughout these years, Stephenson continued to build engines apace, constantly adding refinements as new problems came to light. In 1841 he patented a design for a long-boilered locomotive so as to make better use of the fire tubes as heating surfaces. Locomotives so far usually had a boiler length of about 8 feet 6 inches, but Stephenson's new ones were some 2 feet 9 inches longer. They actually had greater success in continental countries than in Britain because, in order to fit onto the turntables then in use on British railways, the wheels could only span a measurement of 12 feet. Thus they had to be contained in the space between the back of the smokebox and the front of the firebox, resulting in a considerable overhang of the locomotive's bodywork at either end. When hauling freight this was not too much of a disadvantage, but for passenger work, when speeds were greater, it tended to make for rather unsteady riding on the track.

Stephenson was not of course the only engine builder of the time. Another was Edward Bury, who constructed many engines for the London to Birmingham railway. He had been building locomotives for some time and was in fact just too late to compete in the Rainhill trials. In 1831 he produced the 'Liverpool' which was designed as a goods engine and had a firebox with a high-domed crown, earning it the nickname 'Coppernob'. It had four coupled 6-feet wheels and, in contrast to Stephenson's design, two outside cylinders under the smokebox, with the driver's platform attached to the back of the firebox. Bury's locomotives also had a simple bar frame, a design feature that was much adopted by the Americans and became a familiar characteristic of their later locomotives. Unfortunately Bury did not keep up with the increasing requirements of locomotive engines, and he resolutely refused to enlarge the dimensions of his engines or employ more than four wheels. By 1851 they were virtually museum pieces.

Timothy Hackworth, of Rainhill 'Sanspareil' fame, also continued to build locomotives and is sometimes credited with originating the 0–6–0 wheel arrangement. This he introduced on the goods engines he made for the Stockton to Darlington railway. He also produced a special return heating system that ran the length of the boiler twice, ending at the front. For this reason, his engines had a coal tender and a fireman in the front, with the driver and water truck at the back end. One of Hackworth's engines, built for the Stockton to Darlington run, was the famous 'Derwent'.

Some of the most impressive machines of the time were built for Isambard Brunel's Great Western Railway, although considerable problems were encountered initially. In fact of the first twenty locomotives Brunel ordered for the opening of the line in 1838, only two were to prove to be any good. They were both Stephenson engines, 'Morning Star' and 'North Star' – 2–2–2s he had originally built for the $5\frac{1}{2}$-foot gauge of an American railway in New Orleans, but which he converted to the 7-foot gauge. Similar to the 'Patentee' in design, and without the dubious refinements other designers were trying, the 'North Star' was to record a speed of 30 mph hauling a load of 80 tons in 1838.

The name of the engine builder most often

Left: This map shows the railways of England that were in existence by 1852. It can be seen that the basis of the present-day network of railways in England was already established.

Below: Patrick Stirling was the designer-engineer of these 8ft 'Bogie Single' locomotives which ran for many years on the Great Northern Railway. He first produced them in 1870 and a later, improved version ran the first leg of the East Coast journey in the famous Races to the North of 1895.

associated with the locomotives of the Great Western Railway, however, is Daniel Gooch. After the initial failure of the engines, Brunel gave Gooch the task of producing fast, reliable locomotives for the line. Basing his design on 'North Star', Gooch proved himself equal to the task and his fast expresses – notable among them were the 'Great Western', 'Firefly' and 'Ixion' – remained unrivalled for many years. They had a much larger boiler space than hitherto employed, and sixty-two engines were built to their specifications between 1840 and 1842. One of the most famous of the GWR engines was the 'Lord of the Isles', which had 8-feet wheels and was said to run 800,000 miles without being renewed.

In 1840 Britain imported some locomotives designed by an American, William Norris. Used on the Birmingham and Gloucester Railway, a feature of the Norris engine was a four-wheeled swivelling bogie which made for better track holding on steep inclines and sharp curves. Gooch also designed tank engines with trucks or bogies to help to spread the weight. These were to increase in popularity in later years.

Several railways in the 1840s and '50s employed engines of the 'Crampton' design. These inventions of Thomas Russel Crampton were to find greater initial success in continental Europe than in Britain. Crampton was concerned with a problem that was bothering a lot of engine designers, and which Stephenson had attempted to overcome when he produced his long-boilered locomotives – that of lowering the centre of gravity. To this end, Crampton placed the big driving wheels behind the boiler and firebox. The first two engines he built along these lines in 1846 for export to Belgium had 7-foot driving wheels with two carrying wheels spaced out under the boiler. Continental railways were to take nearly three hundred of these engines between 1846 and 1864, many with 8-foot driving wheels. In Britain, some railways were still using them up to 1875.

David Joy's 'Jenny Linds', produced in the 1840s, were a good example of how improvement in actual working had begun to be matched with a greater sense of aestheticism in appearance. The boiler and firebox of the elegant 'Jenny Linds' were covered with mahogany, bound with brass hoops, and copper was used in making the fluted dome and safety-valve covers. Even though the boiler space was small compared to some of Gooch's massive designs, the high steam pressures achieved made them speedy engines. They were developed along 'Patentee' lines with an inside cylinder and a 2–2–2 wheel arrangement with double frames, as also were the 'Sharpies' built by Sharp, Roberts and Company and used by many different railways.

In the 1840s, too, fuel economy was developed by using steam expansively in the cylinders. This was known as the link motion; many people claim it to have been the invention of one of Stephenson's workmen who sold it to his master for £20. In any event, Stephenson was the first designer to use it successfully, although others, such as Gooch, were quick to incorporate the system into their designs. It involved the use of two eccentric rods attached, one at each end, to a curved link. The link was connected by a rod to the reversing hands or wheels in the cab. If the link was lowered as far as possible, the engine would move forward; if it was raised the engine movement was reversed so the engine moved backward. The link could also be used to cut off the steam supply to some extent when the piston had moved only part of the way along the cylinder. The steam therefore became enclosed and was still able to do a considerable amount of work by expansion. Stopping the piston at one-third or half-stroke proved to be sufficient to keep the train moving, except on steep inclines, once it had got underway.

Breaking away from the low-centre-of-gravity theories supported by Crampton and Stephenson, James McConnell built his 'Bloomers' at the beginning of the 1850s. Used on the London and North Western Railway, these 2–2–2s were built to three sizes and had high running plates and inside cylinders. They broke with the 'Patentee' design also in having a pair of inside frames and bearings as opposed to the double outside frames of the earlier engine. The expanse of wheel visible beneath the running plate earned this design of locomotive its nickname, for it somewhat resembled the mode of ladies' dress being advocated by a certain Mrs Amelia Bloomer in 1850!

At the same time as McConnell's 'Bloomers' appeared on the railway scene, John Ramsbotham produced some outside-cylinder engines for the London and North Western Railway. These included the 2–2–2 'Lady of the Lake' engines, and some four-coupled engines known as the 'Precedent' class. Both of these engines were largely rebuilt by Francis Webb. One of the 'Precedent' engines called the 'Charles Dickens' is said to have run every day for twenty years or so between Manchester and London.

From this time on, competition between rival railway companies was forcing them to make the service they offered as efficient and comfortable as possible. Carriages got larger and to cope with the extra weight engineers adopted the bogie designs as developed by Gooch. Its aim was to increase the wheelbase of the locomotives without making them less stable. In fact the bogie principle had been conceived in primitive form as early as 1815 by Chapman and Hedley.

By 1870, locomotives were very handsome and smart affairs. Engineers in Britain, the United States of America and continental Europe were becoming

Top: John Ramsbottom built this class of 2-2-2s for the London and North Western Railway. Originally known as the 'Problem' class, their pleasing appearance and reliable performance earned them the more attractive class name of 'Lady of the Lake'.

Above: In spite of the higher speed and greater safety record of the 7ft wide tracks of the Great Western Railway, the narrow gauge of 4ft 8½in was proclaimed an approved standard by Parliament. This photograph shows one of the broad gauge expresses still in service, although narrow gauge tracks have been laid inside the wide ones.

influenced by one another, although each country retained a few inherent characteristics. The engines produced from now until the end of the century were graceful and externally smooth, and woe betide the driver who didn't keep his engine gleaming and spotless!

Patrick Stirling was one of the famous engineers of the 1870s, although he had in fact been building engines for many years previously. In 1870 he produced the massive 4–2–2 expresses that ran on the Great Northern Railway. Happily iron tracks were now largely replaced by steel, or else they would not have withstood the 15-ton weight of these engines, with their 8-foot driving wheels. William Stroudley, a Scottish contemporary of Stirling, supplied many of the engines on the London, Brighton and South Coast Railway at this time. Noteworthy among them were the 'Gladstones' which had four-coupled wheels in front. Although this arrangement was shunned by many engineers, one of the 'Gladstone' engines was to remain in useful service until 1933. With the change from broad to the narrower (i.e. standard) gauge, Daniel Gooch now had to find a replacement for his 2–2–2s running on the Great Western Railway. He produced the 'Iron Dukes', a class of eight-wheelers. His successor, William Dean, built some virtually identical locomotives but as the 7-foot track was still in use in some places he built some 2–2–2s for the broad gauge, which could be converted to the narrower gauge when necessary.

Designers of locomotives in the United States and in continental Europe were beginning to overtake their British counterparts towards the middle and end of the 1870s, and many of their design features and innovations were copied by British engineers. Greater efficiency was perpetually being called for, together with a requirement to make better use of the steam so as to economize on fuel. Anatole Mallet, a French engineer, successfully employed a principle from marine engineering in locomotive design and produced his compound engine for the Bayonne to Biarritz railway. This incorporated a high- and low-pressure cylinder and the exhaust steam was turned from one into the other to give an extra thrust. Two more Frenchmen, A. G. de Glehn and G. de Bousquet, were to develop and modify this system over the next few years and the engines they built were to become a European type. A. von Borries of Germany used the system to design his two-cylinder compound locomotive which could also be worked as a non-compound.

With British coal both plentiful and cheap, 'compounding' found slightly less favour in Britain, although Francis Webb of 'Precedent' fame was persistent in using it in the engines he designed for the London and North Western Railway. He produced several classes of compound engine – the 'Experiments', the 'Dreadnoughts', the 'Teutonics', and later the 'Greater Britains'. There was some American interest in his compound engines: the Pennsylvania Railroad purchased one, and two went on display at the Chicago World's Fair in 1894.

At this time also there began to be a need for short-distance passenger trains, particularly for the London suburban services. The tank locomotives, which carried their own fuel and water supplies and therefore did not need a separate tender, were particularly useful for shuttle services and a great number of them were made by British railway companies. Dubbed 'the handymen of the line' they were usually equipped with comparatively small wheels so they could get up a fair turn of speed quite quickly, but also had considerable power for pulling heavy loads when necessary. Notable among the tank engines were those designed by William Stroudley; his little 0–6–0 'Terrier' tank engines were well known around London. Stroudley also produced a series of 0–4–2 tank engines for use on the London, Brighton and South Coast railway, and

Top: This 2-4-2 express locomotive built for the Paris, Lyons and Mediterranean Railway replaced the single-drive locomotives designed by Thomas Crampton previously used by this company.

Centre: The swan song of the wide gauge locomotives is marked by one of Daniel Gooch's 'Iron Dukes' as it leaves Paddington for the last time in 1892.

Above: This 0-6-0 'Terrier' tank engine was designed by William Stroudley, specifically for the London suburban services.

Right: Detail taken from John Dobbin's painting "Opening of the First Public Railway – The Stockton and Darlington, 27th September, 1825".

2–4–0 and 4–4–0 tank engines began to replace the single drivers.

The 4–4–0s were particularly popular on the Continent and were later to replace the long-boilered 2–4–2 expresses that had been used on the Paris, Lyons and Mediterranean Railway. Some of the most popular 4–4–0s used in Britain at the end of the nineteenth century were those of the 'Dunalastair' class, produced by John McIntosh. The Belgian State Railways bought five 'Dunalastairs'.

The last years of the nineteenth century also brought another important development towards greater efficiency and economy of steam. This was a

Top: A model of the attractive 4-2-0 locomotive named 'Austria' built by an American, William Norris for an Austrian Railway. Norris also supplied the first American locomotives to run on British railways.

Above: A model of Sir Daniel Gooch's 'Ixion', built for the Great Western Railway. 'Ixion' ran in tests which established the speed supremacy of the broad gauge over the narrow gauge in 1845.

Right: 'Invicta', a sister engine to the 'Rocket' but embodying features of Stephenson's earlier engine, 'Lancashire Witch'. She was built for the Canterbury and Whitstable Railway.

EASTERN
A9
990
90

SOUTH EASTERN & CHATHAM RY. LOCOMOTIVE Nº 737. 1901.

Left above: The Great Northern's 4-4-2, built in 1898 was the first 'Atlantic' type to run in GB. Although they proved entirely satisfactory, they took second place to the large-boilered 'Atlantics' introduced four years later.

Left below: By the 1900s locomotives of the 4-4-0 wheel arrangement had generally been replaced by those with more power. The London and North Western, however, introduced this 4-4-0 at the turn of the century.

Above: One of Francis Webb's long-boilered 'Greater Britain' class which was perhaps more impressive in appearance than in performance. The one pictured here, called the Queen Empress, was later to take part in a publicity campaign for USA's Chicago Exposition.

system that turned into usable steam the water and water vapour inevitably carried and also generated by steam as it travelled. Known as superheating, it involved splitting up the steam by diverting it through a number of small tubes before it went back into a superheated chamber. Subjected to considerable heat from the furnace gases, the steam was dried and reheated so that by the time it reached the cylinders it was nearly double its former temperature and had increased in volume accordingly. Wilhelm Schmidt was the first to utilize the system to its greatest efficiency on the 4–4–0s built for the Prussian State Railway, and gradually, except in France, 'superheaters' began largely to replace the compound engines.

Perhaps one of the most notable events in railway history at the end of the nineteenth century was the fierce inter-company rivalry that found its ultimate expression in the efforts of companies operating the services between London and Scotland. Those involved were the London and North Western Railway, which operated the 'West Coast' route, and the three 'North' companies: that is, the Great Northern, North Eastern and North British Railways, which ran the 'East Coast' route. Initially the LNWR was more renowned for comfort and punctuality, while the East Coasters had the reputation for speed. Towards the end of the 1880s, however, the rivalry intensified into an out-and-out race – the first of the famous Races to the North. Timetables were quickly revised, or even abandoned altogether, as the time for the overall journey was reduced by a whole hour – and still the enthusiastic crews would arrive ahead of the appointed schedule. A far cry from railway operations of today! The first of the Races to the North took place in 1888 and they ceased by mutual consent after the 393 miles from King's Cross to Edinburgh had been clocked in just under $7\frac{1}{2}$ hours.

But the races weren't altogether over! In 1895, with the opening of a direct Highland route to Aberdeen, they began again in even more deadly earnest. This route had been a long time in coming to fruition because of the massive engineering problems entailed in spanning the Tay and Forth estuaries. A bridge had been built across the Tay in 1878, but the following year it was the scene of one of the worst disasters in railway history, when some girders collapsed in a hurricane-force wind, plunging a train carrying 200 passengers into the dark waters below. The sole survivor of the disaster was in fact the train, which was raised from the bottom of the river, repaired and later ran across the new Tay bridge that was completed some years later. The Forth bridge was completed at the beginning of the 1890s. Together these bridges took 17 miles off the East Coast route to Aberdeen.

The 1895 races are possibly even more historic than those of 1888. Even today, trains travelling from London to Aberdeen take longer than the record established during the races some eighty years ago. A variety of engines was used – the East Coasters favouring Patrick Stirling's 4–2–2s and some 4–4–0 engines, the West Coasters for their part using engines of Webb's 'Precedent' and 'Teutonic' classes as well as a famous 4–4–0 known as the Caledonian 'Lambie'. Incidentally, a 'regular' on the Great Northern Railway's London to Edinburgh run was their No 1, otherwise known as the 'Flying Scotsman', which usually took 10½ hours to complete the journey. Throughout the year of 1895, the races gathered momentum with minutes being clipped off the timetables to show earlier estimated times of arrival. Keeping to the schedules meant little to the crews, however, who perpetually endeavoured to arrive ahead of the appointed time: by July, trains were arriving so early that extras had to follow to pick up stranded passengers! A new record would be set up by one company only to be beaten shortly after by the other. The nights of 21 and 22 August 1895 saw the 'final' of the races. The time clocked by the East Coasters on the 21 August was 520 minutes for the 523 miles between London and Aberdeen. On the following night – with the thrill of the chase very much in evidence – the West Coasters recorded a time that is still reputed to be the fastest ever made by rail between the two places: 512 minutes over the 540 miles that comprised their route. Thereafter a truce was called!

To meet the seemingly insatiable demands from all the railways for greater passenger comfort and faster speeds, Harry Ivatt produced the first 4–4–2 'Atlantic'-type locomotive in 1898. Harry Ivatt had succeeded Patrick Stirling on the Great Northern Railway and he was told to produce an engine that could haul expresses weighing 300 tons. A year later Thomas Worsdell, who had left the Great Eastern to join the North Eastern, produced his 4–6–0 express passenger engines, and John Aspinall of the Lancashire and Yorkshire railway produced another 'Atlantic'-type engine. His was slightly more powerful than Ivatt's expresses, but in 1901 Ivatt produced an even larger 'Atlantic', capable of pulling more than 400 tons.

Many new locomotive engineers were to come to the fore in the early years of the twentieth century, as will be seen in Chapter 4. Ironically, however, Britain was to find its locomotive design constricted by the need to conform to the limits imposed by such things as the width of the bridges and height of tunnels already in existence. Many of these had been built to the specifications laid down by Stephenson some eighty or so years before when he was responsible for making Britain a pioneer in the field of railways.

Top: 2-4-0s of the Holden 'T19' class were much favoured by the Great Eastern Railway in the latter part of the nineteenth century – from about 1886. The one pictured here has been re-built – the previously long boiler was replaced by a shorter more squat version.

Above: These little 0-6-0 tank engines are one of the Midland Railway's longest serving engines. They began service in the late 1870s and modified versions were still being used in the 1960s, thus being among the last steam locomotives to be used on British railways. The one pictured here dates from the early 1900s.

Right: The most unusual feature of this 4-2-2 is that it marked a revival of 'single-driver' locomotives for express passenger work on the Midland Railways in the late 1880s. For more than twenty years prior to this, passenger trains on the Midland had all been handled by coupled engines – either 2-4-0s or 4-4-0s.

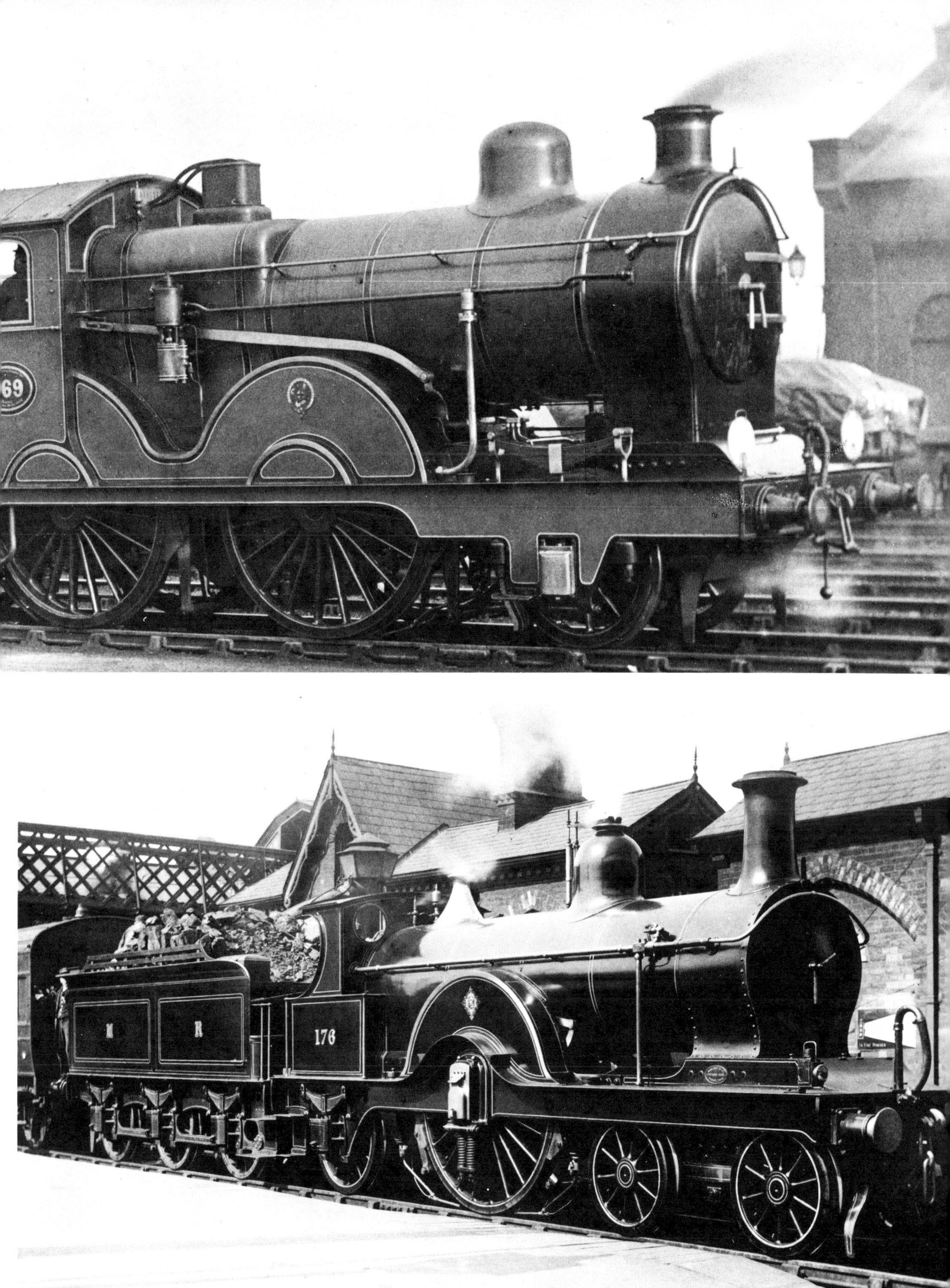
69
M R
176

EARLY AMERICAN RAILROADING

Below: The 'Stourbridge Lion' was the first locomotive to run commercially in the USA, and was put to use on a track previously worked by horses, belonging to the Delaware and Hudson Canal Company at Honesdale, Pennsylvania. Although it worked well, it did not remain in service long as it was too heavy for the insubstantially built track which crossed several flimsy bridges. Opposite below: Design for the 'De Witt Clinton', described below as 'the third Locomotive Engine built for Actual Service on a Rail Road in the United States made for Rail Road between Albany and Schenectady A D 1831'.

The birth and growth of railways and steam locomotives in the United States of America in the early nineteenth century is quite a different story from that of Great Britain. It is fair to say that almost the first roads in the United States except in the New England States were to be railroads, and they were perhaps less to serve a population than to attract one. Whereas in the United Kingdom, the high roads and canals needed assistance in the form of fast and efficient transport to cope with increasing trade, railways were needed in America actually to create trade. Probably for this reason, American railways – or railroads – did not encounter anything like the measure of opposition, even in the early days, and once established, they grew at an enormous rate. In less than fifty years, from 1832 to 1880, the extent of track had grown from just over 200 miles to 100,000.

The first railroad to use a steam locomotive was the Delaware and Hudson; it imported a 'Puffing Billy' type of engine – the 'Stourbridge Lion' – from England in 1829. This was not America's first introduction to steam locomotion, however, for an American by the name of Oliver Evans had done much pioneering work in the field and had demonstrated a model locomotive running along a track some years earlier.

The 'Stourbridge Lion' spurred on a young engineer called Peter Cooper to develop a steam locomotive. A few years later his first engine ran on one of the first public railroads opened for traffic – the Baltimore and Ohio. Called 'The Tom Thumb', it is to American steam locomotion what the 'Rocket' is to British. Doubtless its name was a direct reference to its size, for the 'Tom Thumb' was very small – its four wheels were only 2 feet 6 inches in diameter.

The directors of the Baltimore and Ohio railway were not altogether happy with the 'Tom Thumb' and rightly felt that American engineers could do better, so they offered a prize of $4,000 for the best American-built locomotive to be delivered to them by June 1831. Four engines were submitted, but only one proved satisfactory and that was the 'York', built by Davis and Gartner of York, Pennsylvania. Even this had to be considerably altered before it was really satisfactory and the Baltimore and Ohio Railroad became known for the series of curious-looking engines it was to use for the next few years.

A few months after the Baltimore and Ohio competition, the Mohawk and Hudson Railway – a forerunner of the New York Central Railroad – ran its first steam locomotive. This was the famous four-wheeled 'De Witt Clinton', built by the West Point Foundry. The West Point Foundry also built another famous early locomotive, the 'Best Friend of Charleston', for the Charleston and Hamburg Railroad in South Carolina. Matthias Baldwin, who was to become one of the great names in locomotive building, made his first engine, 'Old Ironsides', for the Philadelphia, Germantown and Norriston Railroad, shortly after the company had seen a model locomotive of his running in 1831.

The 'Tom Thumb', the 'De Witt Clinton', the 'Best Friend of Charleston' and 'Old Ironsides' were all developed largely along the lines of British-designed engines of this time, and the early 1830s in fact saw the import of quite a number of British engines. Predictably, many Stephenson designs found their way to the United States – among them the 'John Bull' for the Camden and Amboy Railroad in 1831; the 'Davy Crockett' for the Saratoga and Schenectady Railroad in 1833, and the 'Stephenson' for the line between Boston and Lowell in 1835. But faced with quite different problems to overcome, dictated by the completely different conditions of the railroads and terrain, American engineers soon came into their own. In fact before long they were virtually leading the world in locomotive design. Copied they may have been by engineers of other

The 4-4-0 wheel arrangement, introduced into the USA in the 1830s, was the forerunner to the later large American locomotives. This one – the General – is one of the most famous 4-4-0s of all time, and owes its fame to its kidnap by James Andrews in 1852 during the Civil War. He used it in an unsuccessful attempt to isolate Chattanooga from the rest of the South.

countries, but no country has been able to rival, or even match, the size of American locomotives.

A feature developed by the American engineers was a swivelling truck or bogie positioned ahead of the driving wheels to help negotiate the curves inherent in many American railroads. It also helped keep the loading light, by spreading the weight of the locomotive as far as possible. This was necessary because the light rails of the railroads were laid on local timbers placed directly onto the ground, making them incapable of bearing great weights. In fact, passenger carriages and freight wagons were also mounted on two bogies, one at either end, to overcome this problem. Pilots or 'cowcatchers' were mounted on the front of the frame at track level, and were an important accessory, for the railroads of the day were mostly unfenced, both in the towns and the country. For this reason too, engines were

equipped with headlights in the front to act as a warning beacon as well as to throw light on the lines. Most engines incorporated a device known as the equalizer beam, which allowed each set of drive wheels to be independently sprung. In order to try and extend the life of the tracks somewhat, the drive wheels were counterbalanced to prevent the rails receiving heavy pounding blows with each power stroke of the piston and rods.

Between 1836 and 1840, the American Standard or the American type engine came into existence. It was a 4–4–0 and the first of its type was built by Henry Campbell for the Philadelphia, Germantown and Norristown Railroad. It became recognized as an ideal design for the rigours of the track when later experiments, such as a 2–2–0 built for the Vermont Central Railway and an 0–8–0 built for the Baltimore and Ohio Railroad proved less successful.

By the 1850s nearly every railroad was using the 4–4–0s and its design was the basis of the massive giants that were to follow.

The number of railroads already mentioned must give some idea of the vast growth of the network. In the 1840s, the most important lines were probably those that were making tracks into the west, namely from Baltimore to St Louis, from Richmond to Memphis and from New York to the Great Lakes. In addition, Chicago and the Gulf of Mexico were linked by the north-to-south line. But besides these major lines, smaller railroads were being developed in isolation at an amazing pace. These lines were constructed out of necessity, to link out-of-the-way settlements, which later had to be built to fit around the railroad. Building of such towns and attracting people to the area by the tracks was ensured by the fact that land adjacent to railroads could be purchased very cheaply in the 1850s.

Because of the nature of the country and the rapid service required, railroad building was not given the careful planning that had been necessary in the much smaller, more overcrowded country of Great Britain. Initially few bridges and tunnels were built. Those that had been were of poor construction: the bridges were liable to collapse under the weight of the trains. Travel on the railroads was a hazardous business and it was something of a gamble as to whether you would reach your destination at all! To make things even more complicated, no specifications or guidelines were issued by the State or Federal governments during this period, with the result that engineers of each line laid tracks to whatever width seemed the most convenient. By the middle of the century, there were no less than twelve different gauges varying from 3 feet to 6 feet.

Along these varying widths of track, however, great speeds of locomotive travel were being reached. 'A mile a minute' became an aim that was both a dream and a nightmare for American engineers as they strove to produce locomotives equal to the task. The first engineer to do so named his engine after himself, and in 1851, the 'Swinburne 71' recorded an average speed of 60 mph over a 34 mile stretch of track near Narrowsburg.

Still there was no railroad that passed beyond the Missouri and the Mississippi. It was almost as if the world stopped there, for the roads in existence beyond were few and inadequate. In 1856, however, the Mississippi was spanned by a bridge, involving a struggle that was to be magnified a hundredfold a few years later. This was the Rock Island Bridge and the opposition to it was so great that a lawyer, Jefferson Davis, was hired to prevent its construction. The lawyer he faced, and to whom he lost the case, was Abraham Lincoln. Five years later these

two men were to face each other in 'battle' again, but this time in much greater earnest – in the American Civil War, with Davis as President of the Southern States and Lincoln as leader of the North. And the railroads were to play an important part in the war, both in the skirmishes leading up to it and in the battles of the day. In fact one of the reasons why the western states fought for the north was to secure the trade connection established by the farmers. Hitherto they had sent their corn to the south by river, for exportation to Europe, but the advent of railroads had made it more convenient to send it by rail to northern ports.

When war actually broke out, the railroads became invaluable for both factions as a fast means of moving large numbers of men and huge supplies of ammunition – both on a scale never before attempted over long distances. An early example was the First Battle of Bull Run in 1861 when the

Top: Railways proved their strategic importance in war, and for the first time ever during the American Civil War, when they were used to transport men and ammunition on a scale that had previously been impossible. The 'General Haupt' was one of the locomotives used for such purposes.

Above: This wood-burning 4-4-0 locomotive – the 'General Sherman' – was used to transport men and construction materials during the building of the Pacific Railroad.

Confederate army under General Beauregard appeared to be losing to the Union men, until support troops suddenly arrived – by train! The following year General Bragg used the railroad to move his entire force from Mississippi to Tennessee. General Sherman was later to say that his historic Atlanta campaign would not have been possible without the railroads. This army of 100,000 men and 35,000 horses received their supplies from four trainloads each day as they marched into the South.

Being of such great importance, railroads obviously became targets both for defence and attack, and entire campaigns were planned around them. Armies from both sides became adept at destroying lines by ripping up the sleepers and burning them and heating the rails until they were pliable enough to wrap round trees. The poor standard of track could almost be said to be an advantage, however, for it proved as easy to repair and re-lay as it was to destroy!

Harking back to the American Standard 4–4–0s, one of the most famous of them all must be the 'General' and it was its capture by a party of Union soldiers led by James Andrews in 1862 that brought it its fame. The train was captured between Atlanta and Chatanooga, the purpose of the mission being to isolate Chatanooga from the South, thus making it vulnerable to attack. This would be accomplished by burning intermediary bridges and destroying telegraph lines. Andrews reckoned without the tenacity and determination of the erstwhile conductor of the train, however, who pursued the captured 'General' in an old handcar, then in a small engine and finally in a big locomotive, the 'Texas'. Andrews meantime was held up at a couple of places by Southern officers and the precious time he had to use to bluff his way through gave the 'Texas' the chance she needed to gain on her quarry. Finally the kidnappers had to abandon the 'General' when she ran out of fuel. By that time she was so hot that the brass parts were melting!

During the Civil War, Abraham Lincoln signed a bill to authorize the first 'trans-continental' railway and telegraph line across the United States. It was in fact to go from the River Missouri to the Pacific coast. Five surveys were made and a route finally chosen that spanned 1,770 miles between Omaha and Sacramento. Two companies were formed: the Central Pacific which was to drive east from Sacramento in California, and the Union Pacific which would lay the line west from Omaha. Although work began at Omaha at the end of 1863, it was not until the middle of 1865 that the first lengths of rail were actually laid. Thereafter two locomotives, 'General Mcpherson' and 'General Sherman', were used to convey men and construction materials. The General himself rode in his namesake in November of that year on an inspection trip. The Union Pacific hired vast numbers of ex-soldiers – many of them Irish – for their part of the line, while the Central Pacific imported thousands of Chinese labourers to work on their sections. The Chinese came to be known as 'Crocker's Pets' after Charles Crocker, who was one of the 'Big Four' directing the Central Pacific.

Never before had railroad construction been carried out on such a huge scale and the problems of

CENTRAL
BAGGAGE
PACIFIC RR
BAGGAGE

Left: A central Pacific locomotive carries the mail across the Sierra Nevada.

Right: Union Pacific's locomotive No 119, built by Rogers Locomotive Works in 1868 was used to pull the special train at the opening ceremony of the Pacific Railroad.

Below: The ceremony at Promontory Point in Utah – when the Union Pacific and Central Pacific Railroads were finally united with a gold spike.

moving men and materials were immense. Central Pacific's first locomotive – 'Jupiter' – had to travel 19,000 miles by sea to California around Cape Horn, and some equipment for the western terminus had to be shipped either on this route or across the Isthmus of Panama. Costs were astronomical, and the Federal government lent money at steadily increasing rates up to $16,000, $32,000 and $48,000 per mile of track. As the Union Pacific blasted its way west, its workers were so rough and disorderly that their camps, which resembled towns, came to be known as 'Hells-on-Wheels'. During the course of their work, they were attacked by tribes of Indians who resented the intrusion of the 'iron horse' across their lands. Peace was finally made with the Indians in two ways: General Grant, who had commanded the Northern forces during the Civil War, smoked a pipe of peace with the chiefs; and, probably even more effectively, the once fierce warriors were seduced by the introduction of alcohol. The Chinese on the Central Pacific for their part had most of their battles with the terrain and elements of nature. Rigours to be faced included blasting tunnels out of the granite rocks of the Sierra Nevada ranges, while combating the freezing snowfalls at the same time.

The whole venture became a fierce battle between the two companies, who competed at one stage to see who could lay the greatest length of track in a day. This was no friendly rivalry, however, but deadly earnest combat, with a $10,000 reward for the winner. When the Chinese won this particular test of speed by laying just over ten miles in one twelve-hour stretch, it did nothing to improve relations between the two factions. Doubtless the quality of the track suffered with this emphasis on speed, but that seemed to bother the promoters little. Their main aim and anxiety was to collect the bonds paid as each mile was laid – and so the faster the better. In addition the more track each company laid, the more land adjacent to the track it would have to sell at a vast profit.

The building of the line was a revelation to the Europeans who were used to railways being scientifically engineered and carefully constructed from iron, stone and cement. Not so the Pacific Railroad – where earth replaced more substantial material as ballast and rough-hewn timber was used for the ties.

Even more amazing, and absurd, was what happened when the rival companies lines actually met – both sides just went on building! Over 200 miles of extra track was laid, the rails running parallel, destined, it seemed, never to meet! At this stage the antagonism between the workers manifested itself in downright aggression as the gangs failed to warn each other when they were planning to blast, thus showering each other with rocks and stones. Finally Congress stepped in and decided on the meeting point for the rails. This was to be Promontory Point, in Utah, and the famous 'gold spike' ceremony took place on 10 May 1869. The first engines, Central Pacific's 'Jupiter' and Union Pacific's 'No 119', had been driven along the line from west and east until their cowcatchers touched each other. Then President Leland Stanford of the Central Pacific united the rails with a gold spike driven in place with a silver hammer. As the engineers of the locomotives broke a bottle of champagne on each other's engines, news of the completion of the historic line was sent buzzing to the nation along the new telegraph line.

The Pacific Railroad marked the beginning of the railroad building in the West, but it wasn't to be a lone pioneer for long. One thing it achieved,

however, was more or less to establish a standard gauge, for Congress had stipulated the tracks laid down by both companies must be 4 feet 8½ inches. This was the gauge that almost all new railroads were to adopt in the future, although some discrepancies in the track remained for some time. The Erie Railroad for example retained its wider 6-foot gauge until 1878 and 5-foot gauges remained on many lines in the South until 1886. An amazing variance of just ½ inch was in existence on some lines where the tracks had been laid 4 feet 9 inches apart. The Louisville and Nasheville was one example, and many railroads kept this width until 1900 and even later.

The two historic locomotives – 'Jupiter' and 'No 119' – that took part in the opening ceremony of the Pacific Railroad deserve a further mention. Both were 4–4–0s, the most popular locomotive in America then and for at least another decade; 'Jupiter' was built by Schenectady in 1868 and 'No 119' by Rogers the same year. Designed for freight or passenger service they could both pull about twenty freight cars at 10 mph or five to six passenger carriages at 25-30 mph. The main difference in their appearance was in their stacks: 'Jupiter's' was characteristically funnel-shaped; 'No 119' had a tall straight stack. Their brightly coloured shiny paintwork, gleaming brass and iron fittings and gilt lettering decorations characterize them as typical locomotives of the 1850–70 period. It was a reflection of the 'rococo' art of the time and many locomotives bore brightly painted landscapes, animals or portraits surrounded by elaborate 'scrolls'. This extravagant appearance replaced the utilitarian, unromantic look of earlier locomotives, just as it would soon give way to the more practical demands of later years.

Opposite: A dramatic shot of an American Standard 4-4-0 locomotive crossing a crudely constructed wooden trestle bridge. Built in the 1860s this is a wood-burning locomotive as can be seen by the stacked planks in the tender.

Above: This example of a Fairlie type locomotive was ordered by the Denver and Rio Grande in the 1870s. Typical of its type it comprises two engines connected back to back.

Already the immense distances that locomotives had to cover had led American engineers to design engines with greater fuel and water capacity and more tractive power. The Baltimore and Ohio Railroad by then had ordered some 4–6–0s to assist their 4–4–0s. The 4–6–0 was a huge and heavy locomotive, considered to be a prestige model and often used to pull the 'hotel' carriages introduced by George Pullman. Pullman's carriage, designed with facing seats that could be made into beds, had first been used in 1859 on the lines between Bloomington and Chicago. The real ancestor of his 'luxury' coaches, which were to become world famous, was 'Pioneer', built in 1863. America in fact was to have open passenger cars and 'sleeper' carriages long before Great Britain, where the first Pullman coach was not seen until 1873.

Also in the 1850s, the Lehigh Valley Railroad introduced the huge 2–8–0s specifically for hauling wagons filled with anthracite coal. The first engine was called 'Consolidation', which came to be the name applied to the 2–8–0– wheel arrangement. They were really a natural development of the successful 2–6–0– 'Mogul'-type engines which had such an impressive record on the Pennsylvania Railroad for fast, heavy freight service. This design was copied in Great Britain in 1873.

Although the Lehigh Valley Railroad first introduced the 'Consolidations' in 1866, it was almost another decade before they enjoyed widespread use. Capable of hauling 1,000-ton freight trains at 14 mph, the 'Consolidations' marked the end of the dual-service, freight-and-passenger, locomotives, for after their introduction, the 4–4–0s were almost entirely for passenger service. On the Erie Railroad it was estimated that one 2–8–0 could almost do the work of two 4–4–0s, and a total of 33,000 'Consolidation'-type engines were said to have been made for America. American designs for huge engines did not stop there, however, for a year after the 'Consolidations' first appeared on the rails of the Lehigh Valley Railroad, that same company ordered its first ten-coupled engine, called 'Decapod'.

An important advance that came with the development of these heavy locomotives was the replacing of iron rails by steel. Had this not taken place, the heavy engines would not have been possible, for iron rails would never have been able to sustain their weight.

Following on from the building of the important Pacific Railroad, construction of many long and important lines began to get under way. The 'Railway Mania' echoed that of the 1840s in Britain with the competition among the railway men, or 'robber barons' as they came to be known, every bit as fierce and cut-throat. The influence of these men on the nation and its wealth was enormous and in effect they dominated the financial life of America for nearly fifty years.

The year 1869, which saw the completion of the Pacific Railroad also brought the beginnings of the Atchison, Topeka and Santa Fé Railroad, later to be incorporated into the Chicago-California Santa Fé system. A new type of engine was introduced on the Atchison, Topeka and Santa Fé Railroad, known as the 'Santa Fé' and the first one to appear had two outside cylinders arranged in a line on either side. The front ones took steam from the boiler and passed it on for re-use to the larger cylinders behind. The pistons that worked the two cylinders were located on the same piston rod, which extended through each pair of cylinders. This arrangement was known as tandem compound.

The Santa Fé Railroad blocked the proposed route of a line chartered by the citizens of Denver in 1870 to link them with Mexico. This was known as the D & RGW – The Denver and Rio Grande Western. Later it was to push its tracks through the Colorado Rockies after the discovery of gold and silver in the mountains. It was one of the few to be granted an exception to the standard gauge, for 3 feet was found to be more suitable for the mountainous terrain.

Also in 1870 the Northern Pacific began pioneering into the Pacific Northwest and the St Paul and Pacific Railroad – a former venture that had faded – was revived to become the Grand Northern. The Pacific Northwest was completed in 1883, and the Grand Northern reached Seattle ten years later in 1893.

The 'Big Four' of the Central Pacific – that is Charles Crocker, Collis Huntingdon, Leland Stanford and Marc Hopkins – formed the Southern Pacific which pushed inland through the Southwest to New Orleans. The Big Four were among the leaders of the influential, and often corrupt, railroad men, and they also established their own construction companies to whom they awarded individual and highly lucrative contracts. Making personal fortunes for themselves, this practice left the railroads as such with huge debts, bankrupting its other promoters and doing little to enhance the reputation of the industry.

Besides the frantic railroad activity in the West, the Middle West too became laced with railway network as immigrants poured in to exploit the rich agricultural land. Notable railroads were the Burlington, Milwaukee, Rock Island and Chicago and North Western each incorporating several branch lines as well as their main lines.

A key figure of the railroads at this time was Cornelius Vanderbilt, who as a young boy had helped his father with his business on the waterways.

The ships that served the inland waterways of eastern America experienced great problems in the severe, snowbound winters. Vanderbilt was quick to see how their service could be replaced by the railroads, which were less vulnerable to and less limited by weather conditions. He bought major shareholdings in railroad companies. This wealth led him to become the guiding influence behind the Hudson line and the New York Chicago, as well as the New York Central System. His battle with Dan Drew for control of the Erie is one of the well-known fights of the time.

The railroad scene attracted other 'giants' such as Jay Could, who became President of the Erie and did much to set Union Pacific on the potential road to ruin. UP was later bought by another man, Edward Harriman, who had become a director of the Illinois Central. He then mortgaged Union Pacific to raise the necessary capital to buy and improve the Southern Pacific.

The public, however, were beginning to raise their voices against these crippling, monopolistic railroad practices, for they of course were the victims of the 'rate wars' that led to both artificially high and discriminatory rates. As a result the Interstate Commerce Commission was established by the Federal government in 1887. Over a period, it exercised increasing control over the activities of all railroads that crossed State boundaries. Breaks at rivers and cities began to be eliminated and rolling stock gradually became standardized so that it could move more freely over longer distances. This of course helped to put a stop to the self-motivated practices of the 'robber barons' for the small railroads of purely local importance were the only ones that did not come under the Commission's jurisdiction. They were small fish to fry after the previous huge, megalomaniac ventures!

Greater control over the standardization and general organization of engines and locomotives had come into existence before the control over the railroads, however. It had been helped by the founding of the American Railway Master Mechanics Association, among whose main objectives was the determination to standardize motive power.

A notable advance incorporated into locomotive building during this era was George Westinghouse's air-brake system, which he perfected in 1872. It brought a great step forward in train operation safety for hitherto each carriage had to be slowed manually by individual brakesmen. The Westinghouse brake worked on a system of compressed air relayed from the engine to all carriages through air pipes. These were connected to one another by

Below left and right: Two 4-6-0 locomotives of the Chicago, Milwaukee, St. Paul railway photographed at the turn of the century.

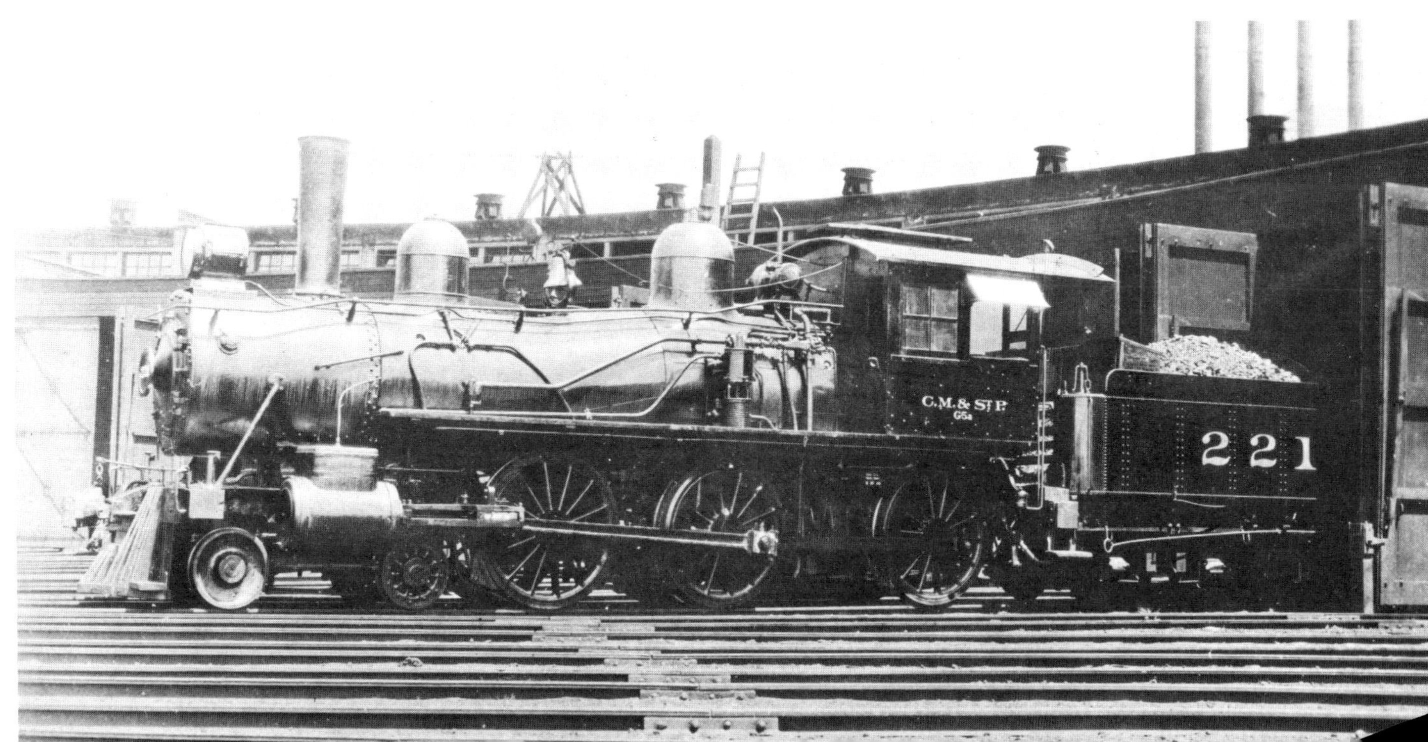

flexible couplings. As long as air pressure was maintained in the pipes the brakes were held 'off', but if air was allowed to escape – either because the driver operated the valve, or in the event of carriages becoming uncoupled – it passed to the brake cylinders thus bringing them into action simultaneously. It could be applied to all types of engines and rolling stock and was immediately adopted by all the American locomotive builders as well as many in other countries.

It was followed some years later by another safety feature. This was the Master Car Builders' automatic car coupler that could be used to link all the cars and carriages in a train together. Like the Westinghouse brake, it was far in advance of any other form of coupler hitherto used and it replaced the somewhat dubious link-and-pin system in use before it.

In the late 1870s the Wootten firebox was designed by John Wootten of the Philadelphia and Reading Railroad. It could be applied to nearly all the wheel arrangements in existence and was very shallow and wide, thus providing the large grate area needed for the slow-burning anthracite used by many railroads. Wood and anthracite coal were the fuel used in all early locomotives, with wood predominating initially as it was plentiful and left few ashes. The opening of the bituminous mines, once the greater heating properties of bituminous coal had been realized, led many companies to use this as their main source of fuel. Both wood and anthracite were relatively clean fuels, however, and were often retained for this reason.

When anthracite, which burns slowly with little smoke, was first used it was obtainable in fairly large sizes which brought few problems. The supply-and-demand principle, however, meant that it became more expensive and increasingly smaller grades had to be used. These were difficult to keep fired in the small fireboxes on most locomotives and needed a much larger grate area to produce the same total amount of heat release given out from smaller grates burning bituminous coal. Which is why, of course, John Wootten designed his wide firebox! It was suitable for the slack and virtual dust of anthracite waste coal, as well as other inferior grades of coal. As the Wootten firebox was placed above rather than inside the rear drive wheels, this meant there was no longer room for the driver to go in his usual place alongside and behind the firebox. His cab was therefore moved forward and positioned ahead of the firebox.

The Wootten firebox was to have a distinct influence on future design, particularly on the deep, wide firebox that was to be introduced for burning bituminous coal. One of the first locomotives to have this incorporated into its design was the experimental 2–4–2 type built by Baldwin for the Chicago, Burlington and Quincy Railroad in 1895. The firebox was placed behind the four coupled driving wheels and over the trailing wheels. This allowed for larger driving wheels and deeper fireboxes to be used without placing the boiler at an excessive height above the rails.

Soon after the successful introduction of the 2–4–2s, the State Railways of Japan asked Baldwin to design a freight engine for use on the Tokyo to Yokohama Line. Baldwin produced the famous 2–8–2 'Mikados' or 'Mikes' which pleased the Japanese and also found instant success in the United States. Used by all major railroads, the 'Mikados' were the predominant freight engines for a couple of decades.

During the final years of the nineteenth century, America could boast the largest and most powerful locomotives as well as the fastest. One of the best-known 4–4–0s, 'No 999', recorded a memorable speed of 112.5 mph. But besides the prowess of speed, there was beginning to be a need for passenger engines with greater pulling power than the 4–4–0s could provide. The Baldwin works came to the fore yet again, when it produced the 4–4–2s to meet the Philadelphia and Reading Railroad's need for fast but powerful engines for its 'Atlantic City Fliers'. As we have already seen, the trailing-wheel arrangement allowed for the provision of a larger firebox and thus more steam could be produced. The driver, in his newly positioned cab midway over the boiler and ahead of the firebox, kept in touch with the fireman by telephone! Superbly reliable engines, these 'Atlantics' as they were called, hauled heavy loads regularly and were soon used by most railroads.

With this success behind it the company went on to design the even bigger 'Pacifics' – 4–6–2 initially produced for New Zealand. This was another export that found immediate success on home ground too, and 'Pacifics' were to become the recognized passenger engines in the same way that freight travel had become synonymous with the name 'Mikado'. Four years into the twentieth century, the Baldwin company produced a 160-ton 'Pacific' for the Union Pacific. With coupled drive wheels of over 6-foot diameter, the manufacturers claimed it as the largest and heaviest locomotive in the world, a distinction it would not hold for long.

Right: The 2-8-0 'Consolidation' type was introduced into the USA in the 1860s by the Lehigh Valley Railroad for freight work, and the class became very popular for heavy work throughout the USA as well as other parts of the world. The 2-8-0 pictured here was retained at Englehart, Ontario, long after steam locomotive power had been largely replaced on the railways, to run trips for steam enthusiasts.

Overleaf: A model of a typical American locomotive of the 1870s.

137
137
137

Wm R. LENDRUM
BUILDER
SCRANTON, Pa. U.S.A.

Left above: A 2-6-2 built by the Baldwin Locomotive Works in 1913. Locomotives of this wheel arrangement were known collectively as 'Prairies' and this one is unusual in that it is still burning wood as its fuel. By 1880, ninety per cent of American locomotives had changed to using coal, following the opening of several bituminous mines.

Left below: A contemporary photograph of one of the famous 2-8-2 'Mikados' first built by the Baldwin Locomotive Co at the end of the nineteenth century for export to Japan. This locomotive found instant favour in the USA as well and is seen here pulling an excursion train over the 'Rat-hole' line at King's Mountain in Kentucky. The abandoned tunnel is a result of constructional improvements made on the line in the 1950s.

Above: Two locomotives revived to run fan trips, doublehead on a Southern Railway run from Fort Valley to Atlanta. They are a 4-6-2 'Pacific' type, first built in the USA at the beginning of the twentieth century and a 2-8-0 'Consolidation' type. Doubleheading is necessary either when the track involves a very steep incline or if a long, heavy train is to be hauled.

Right: The 4-6-2 'Pacifics' became the most popular and widely used passenger train engines in the USA soon after their introduction. No 5629 pictured here, was originally owned by the Grand Trunk Western, and was later bought privately by a steam locomotive enthusiast.

THE GREAT STEAM LOCOMOTIVES OF THE 20th CENTURY in America

At the beginning of the twentieth century there were approximately 200,000 miles of track throughout the United States with an estimated figure of 40,000 steam locomotives running on them. In spite of the terrific production of locomotives in the latter years of the nineteenth century, it had become evident that steam locomotion still had to be much improved in terms of increased power if it was going to keep pace with the ever-growing demands being put upon it. The two-wheeler trailer-truck engine had been universally adopted and had gone far in helping produce more powerful locomotives, but engineers were busy developing

Right above: One of the 'Berkshires' used by the Nickel Plate Railroad on its main line from Buffalo, New York to Chicago, Illinois and St Louis, Missouri. Nickel Plate's fleet of 'Berkshires' were considered to be the finest examples of the class.

Right below: One of the 2-8-4 'Berkshires', the first versions of which were built by Lima in the mid 1920s. Lima originally christened these locomotives 'Super Power', but they became known as 'Berkshires' when they were put into service by the Boston and Albany in New England's Berkshire mountains. The one pictured here was built in 1944, soon after which it was transferred to a museum as a show piece. However, it was brought back into active service in 1968 to run fan trips.

Below: The Western Maryland Railway brought these 2-8-0s into freight service in 1914, at a time when most railroads were using much larger, articulated locomotives for freight. They were, however, one of the most highly developed of their class in the world.

further refinements. One of these was a mechanical stoker which was incorporated into engine design and enabled the hungry firebox to be fed continually without pushing the physical strength of the fireman to its limits. Superheating, as discussed on page 25, was being adopted by all locomotive builders and would do much to increase power output. The replacement of iron castings by ones made of steel, which was both lighter and stronger, made it possible to build still larger engines.

The cries from the railroads for bigger, more powerful engines could not be satisfied by turning out just a few engines, however. New locomotives were needed quickly all over the country, and in far greater numbers than ever before. To help increase the rate of production and bring in some measure of standardization, in 1901 eight locomotive-building plants merged into one new organization: The American Locomotive Company, known as Alco. Its main competitors in the field were the Lima Locomotive Works, and the well-known Baldwin Locomotive Works, which had built its first engine in 1831 and had recorded an impressive output ever since. Although Lima was a much smaller concern that the other two, it is recognized as one of the three chief locomotive builders of the early twentieth century. These three companies were known as the 'Big Three' throughout the United States.

Some of the first engines to emerge in the early 1900s were the Mallet compound articulated engines. Anatole Mallet had devised the compound system some twenty years earlier (see Chapter 1), and his articulated compound locomotives had been introduced in Europe towards the end of the nineteenth century. The principle of the Mallet articulated engine was the use of two sets of coupled wheels on which the boiler was mounted. The back set was rigidly attached to the main frame while the front set was only jointed to it. The front wheels were therefore able to swing or swivel to right or left to follow a curve in the track. The compound system worked in such a way that the steam was first used to drive the rear-coupled wheels before being relayed forwards for use again at a lower pressure to drive the front set. The manoeuvrability of these locomotives made them ideal for lines with sharp curves and severe gradients – frequent features of American railroads!

The 4-6-4 'Hudson' class was pioneered in the USA by the New York Central at the end of the 1920s. Used for heavy express passenger service, they soon proved to have a much greater haulage capacity than the 'Pacifics' which had been the most popular passenger engine up to this time. The first models were built by Alco who went on to improve and develop the design for the next ten years or so.

NEW YORK CENTRAL
5445

The articulated engines were first built in the United States by Alco who produced an 0–6–6–0 type for the Baltimore and Ohio Railroad. Two years later in 1906, Baldwin produced its 2–6–6–2 for the Great Northern Railway. These locomotives, with their leading and trailing trucks or bogies, were better adapted to American track conditions than their predecessors and proved to be one of the most popular of all articulated types. The following year Alco brought out its first 0–8–8–0 for the Erie Railroad. Enormously heavy engines, they were used solely for heavy grade work, but they were topped a few years later in 1909 by Baldwin 2–8–8–2s built for the Southern Pacific Railroad. Five years later, the Baldwin company produced the triple articulated Mallet compound, the 2–8–8–8–2 needed by the Erie Railroad to cope with the ever-increasing length of its freight trains. Although this locomotive had three engine units, it still worked on the compound principle, with the high-pressure cylinders driving the middle group of wheels and still the only ones to take steam direct from the boiler. It was then piped for re-use into two cylinders attached to the front wheels and two attached to the rear wheels. One of the 2–8–8–8–2s is claimed to hold the all-time record for the heaviest train hauled by a single engine. It pulled 250 freight cars which stretched just over 1½ miles and weighed 15,300 tons.

Railroads that did not favour the articulated engines turned to the 2–10–0 and 2–10–2 for their heavy work. The 2–10–0 type had originally been built as an experimental engine towards the end of the nineteenth century, but the little use it found indicated it was not a great success. At the beginning of the twentieth century, the Atchison, Topeka and Santa Fé revived the 2–10–0 with two huge engines fitted with tandem compound cylinders. This was almost immediately followed by the 2–10–2, which retained the tandem compound cylinders, but took over the title from the 2–10–0, of the 'world's heaviest steam locomotive'. It would soon relinquish this, however, to the heavier articulated engines.

By 1910, the 4–6–2 'Pacifics' were well established as the most popular passenger engines, and in July of that year, Alco built an experimental 4–6–2, 'No 50,000', which was probably the most powerful engine of its type in existence. As an experiment, it was clearly successful, for this engine was still in daily use on the Erie Railroad some thirty-five years later! The following year, 1911, Alco added an extra pair of driving wheels to the 4–6–2 design to produce the 4–8–2 'Mountain' type. This wheel arrangement became established as a standard for heavy passenger service.

In 1918, with America now involved in World War I, the government formed the United States Railroad Administration which took over operation of the railroads. With the builders now required to produce locomotives and heavy armament for military use, representatives from the three main companies met to approve twelve designs in eight different wheel arrangements. The locomotives built to these specifications were known as the USRA engines and they represented nationwide practice. Such standardization was too good to last of course,

Above: Union Pacific's No 8444, which was revived to run fan trips. It is the largest steam locomotive to be used for this purpose.

Opposite above: Alco's developments to the 4-6-4 resulted finally in the streamlined version produced in 1938. Known as the 'Twentieth Century Limited' it was an outstanding success and achieved a great reputation for punctuality in service.

Opposite below: The claim to fame of Union Pacific's locomotive No 8444 is that it was the company's last operating steam locomotive. It was a 4-8-4 built by the American Locomotive Company in 1945 – a time when the diesel-electric locomotive was already making rapid headway in America. They were considered magnificent locomotives and Alco's aim in building them was to achieve with steam all that was currently being claimed for diesel.

and indeed it disappeared almost immediately when the railroads had their properties returned to them in 1920. There followed a five-year period of adjustment by the railroads and the locomotive builders when few new types of engine were developed, although work was carried out on improving existing models. For example, a one-piece engine-bed frame was developed to replace the bolted-assembly, bar-frame structures, and articulated engines using simple cylinders instead of the compound arrangement were produced.

By 1925, perhaps because of the lack of new designs over the preceding five years, the railroads were not keeping up with the demands of the economy. Particularly in freight transport there was a need for faster schedules and more efficient locomotives. In an attempt to prove itself equal to the problem, the Lima Locomotive Works produced a new type of engine – the A1, which had a 2–8–4 wheel arrangement. The four-wheel trailing truck had become necessary to support the huge firebox, which in turn was needed to ensure that steam could be produced at the capacity required to sustain the increased power. In addition, the A1 had a 'booster' device in the form of a small engine located in the trailing truck beneath the locomotive cab. Steam was fed through this to engage a gear, which then meshed with another gear connected to the trailing-truck axles. The extra turning power this gave to the small wheels gave additional traction to help in moving a heavy stationary train.

Lima built this engine for the Boston and Albany Railroad which ran through the Berkshire mountains of New England. Christened the 'Super Power' by its builders, the 2–8–4 type soon became generally known as the 'Berkshire' and in various forms it was built for many railroads for several years to come. Some of the finest of all were the 700 class 2–8–4s built for the Nickel Plate Railroad and used on its main line from New York to Illinois and Missouri. These superb trains enabled the Nickel Plate Railroad to compete successfully with the much bigger New York Central System.

In the same year that Lima introduced its 2–8–4s, Alco brought out its first three-cylinder 4–10–2 type for the Union Pacific, closely followed by similar, but rather larger engines for the Southern Pacific. In 1926, Alco again incorporated the three-cylinder design in a 2–12–2 type, but it is a year that may be particularly associated with the Baldwin Locomotive Works for it produced its 607,000th engine – a three-cylinder compound of the 4–10–2 type.

Most locomotives from 1926 onwards incorporated the four-wheeler trailer truck introduced on the 2–8–4s. These locomotives had been built mainly for freight operations, and manufacturers were quick to bring out a corresponding locomotive designed for passenger use. This was the 4–6–4

'Hudson' type, the first and most famous of which were the J-1 engines built by Alco for the New York Central. In effect the 4–6–4 was a logical extension of the popular 4–6–2 'Pacifics', but given more steam-producing capability by the large firebox, and thus able to maintain the fast speeds while pulling larger and heavier trains. The first 'Hudson' delivered to the NYC, the 'No 5200' pulled a twenty-six car train weighing well over 1,500 tons at 75 mph. Shortly after this, the railroad's chief engineer, Paul Kiefer, introduced a dynamic counterbalancing modification to the wheels of all NYC's 4–6–4s. Thereafter, led by the famous '20th Century Limited', the 4–6–4 'Hudsons' gained fame by pulling the passenger trains of the NYC 'Great Steel Fleet'. Some years later Alco improved the 4–6–4 still further for the New York Central by producing the J-3 class, or 'Super Hudsons'. These had even more hauling capacity, but due to improvements in design cost no more to operate. Some of these engines were covered by a streamlined casing, and they were equipped with one of the earliest complete applications of roller bearings on the main side rods. Tested against the old 'Hudsons', they showed their prowess by producing 10 per cent more boiler horsepower and over 20 per cent more cylinder horsepower, while actually burning less coal.

Before the 'Super Hudsons' 4–6–4 J-3 class, however, came the mighty 4–8–4s in answer to the railroads' request for a locomotive that could be used for either freight or passenger work. Not for nearly fifty years, since the American Standard 4–4–0s reigned supreme, had there been a locomotive that could be used interchangeably. The first 4–8–4s were built for the Northern Pacific in 1927 – and here at last was a locomotive that was equally at home hauling passenger carriages or freight cars. Not surprisingly other railroads were quick to incorporate them into their service and a locomotive type that had been known as the 'Northern' was soon as well known by such names as the 'Niagara', 'Dixie' and 'Golden State', as its use became widespread. The last steam train ever ordered by the Union Pacific was a 4–8–4, 'No 8444' in 1944 – the last also of forty-five 800-class 'Northerns'. It was used for hauling the mail-and-express trains and the secondary passenger trains of UP's Omaha to Los

Opposite: A 4-8-4 built and owned by the Reading Company in 1945. 4-8-4s were also built by Alco (see pages 50 and 51), but it was those built by the Baldwin Locomotive Company which were considered among the most magnificent of any built in the world.

Below: A Union Pacific Mikado locomotive at La Salle, Colorado.

Angeles and Portland service. Many years after diesels had come into operation, UP overhauled 'No 8444' to perform special excursion trips. The steam enthusiasts who have ridden the excursion run are quick to point out that not only does 'No 8444' regularly clock 90 mph, but it has not infrequently been known to take the imposing gradient of Sherman Hill in less time than it takes the modern streamliners pulled by five diesel units!

Back however, to the beginning of the 4–8–4s, when the purrings of diesel were little more than a slight thorn in the massive flanks of the steam locomotive. At this time, Baldwin built some 4–8–4s for the Atchison, Topeka and Santa Fé Railroad that were claimed as the largest locomotives of their type in the world. They were impressive engines indeed, again able to pull loaded passenger or freight trains with equal ease. For the same railroad Baldwin followed them with the mighty 2–10–4 locomotives, built exclusively for freight. These giants had drive wheels 74 inches in diameter, larger than those of many passenger locomotives of the day.

Following on from the 4–8–4s was the first four-cylinder articulated engine to have a four-wheel trailing truck. This was 2–8–8–4 built for the Chicago, Rock Island and Pacific. Almost immediately after its inception, Baldwin produced the first 2–8–8–4 to operate with the cab end at the front. It was designed specifically for the Southern Pacific whose lines had many tunnels and snowsheds.

The ultimate expression of power and might in the articulated locomotives must be the famous 'Challenger' 4–6–6–4s, built in 1936 and followed five years later by the 'Big Boys' – 4–8–8–4s. Both built for the Union Pacific by Alco, they were both immediately and undeniably an outstanding success. 'Big Boys' – new contenders for the 'largest engines in the world' title – were built specifically to meet the railroad's request for a locomotive powerful enough to take freight trains over the Wahsatch mountains without the assistance of a helper locomotive. Its nickname was established when a workman, so impressed by the monolithic machine that confronted him on the erecting floor, chalked the name 'Big Boy' on its side. Other impressive Mallet articulated locomotives were built at the same time, including the giant 2–8–8–8–2s, known as the 'Triplex Mallets' and built by Baldwins for the Erie Railroad. These massive engines, of which only very few were built, were capable of hauling a load of over 15,000 tons and it wasn't an unusual sight to see one at the head of a freight train some four miles long. Between them all the Mallet locomotives must surely be claimed as the most tenacious competitors of the new forms of tractive power.

During the decade of 1940–50, the diesel-electric locomotive was beginning to make its influence felt fairly convincingly on the steam-locomotive industry, and few new steam engines were brought out. A logical development of the steam locomotive was introduced in 1944 by the Baldwin works in

conjunction with the Westinghouse Electric Company: a 6–8–6 engine for the Pennsylvania Railroad, powered by a non-condensing steam turbine with a geared transmission to the drivers. Three years later, Baldwin acquired the Lima Locomotive Works and formed the Baldwin-Lima-Hamilton Corporation. Involved in making a few engines for the home market, the company's main concern now was to make steam locomotives for export, for 1950 clearly marked the end of the steam engine's dominance in the United States. Railroad motive power was to shift to other forms.

Left: One of the largest steam locomotives ever built – the articulated 4-8-8-4 'Big Boy' built by Alco for the Union Pacific and brought into operation in the closing years of steam operation. The railroad put twenty into service in 1941 and five more three years later in 1944. Although they were initially designed for use across the Wahsatch mountain range, UP took to using them on the Wyoming division of the railroad.

Above: A Baldwin-built 2-6-6-2 'Mallet' Articulated Compound Locomotive blowing steam from its cylinders. Mallet articulateds were built with varying wheel arrangements but they were all based on a design introduced by the French engineer, Anatole Mallet. In spite of their huge size, they were suitable for use on badly laid lines with severe gradients and sharply curving track.

THE GREAT STEAM LOCOMOTIVES OF THE 20th CENTURY in Britain

As British railways moved into the twentieth century, rail travel was undoubtedly the fastest, most efficient and therefore the most important form of land transport. It enjoyed this pride of position the world over, encountering little competition, particularly for journeys outside urban areas. Even though a few more decades would see the end of the railway's monopoly, there was as yet no indication of it to cast a gloomy cloud on the horizon; petrol-driven vehicles represented no immediate threat and in some large cities, underground rail systems were even ousting the electrically powered, street-level trams.

With such an eminent position to maintain, there was a never-silent demand for larger, more powerful locomotives: a demand that British engineers were not slow in meeting. Some railways, although not all, began to use 4–6–0s and the 4–4–2 'Atlantics' such as those produced by Ivatt, Aspinall and Worsdell and mentioned in Chapter 1. Four-cylinder

locomotives were also developed. George Churchward, who had been appointed locomotive superintendent of the Great Western Railway, in the early 1900s met the demand initially by importing three four-cylinder compound 'Atlantic' engines from France for trials. They were called 'La France', 'Alliance' and 'President', and to match against 'La France' in the trials, Churchward used his two-cylinder 4–6–0 called 'Albion'. It was subsequently changed to a 4–4–2 and compared favourably to the French engine. Churchward then modified the 'Albion' still further by installing four cylinders instead of two. He renamed it after that great classic of the Great Western Railway 'North Star'. It was to prove so successful that it was rebuilt to a 4–6–0 and became the first of many 'Star' engines to run on this line. One of the later 4–6–0 'Star' engines, christened 'King Edward VIII' was used to convey the funeral cortège of its royal namesake from Paddington to Windsor. Churchward was certainly one of the founder members of the 'modern' school of locomotive practice and his designs were quite new, if not revolutionary. He dispensed with the dome and introduced a tapered boiler barrel with top feed. From his 4–6–0 engines he went on to design the first, and for many years the only, 4–6–2 'Pacific' locomotive. The 'Pacifics' had evolved from the wide-firebox 'Atlantics' that had been developed in the USA to meet the need for a larger firebox and grate to help increase the production of steam. 'Pacifics' were introduced into France and Germany before 1910 and Churchward's 'Pacific', known as 'The Great Bear', took to the rails of the Great Western Railway in 1908. It proved to be almost too in advance of its time as other engineers of the line could not reconcile themselves to such a huge machine – it weighed $97\frac{1}{2}$ tons – pounding along the lines west of Bristol. Several years later it was rebuilt as a 4–6–0.

Two notable designers of the decade 1900–10 were

Left: This 0-6-0, built in 1901, was the standard design used for freight operation for many years by the South Eastern and Chatham Railway. This specific locomotive is now owned and operated by the Bluebell Railway in Sussex.

Above: One of the French-built four-cylinder 4-4-2 'Atlantic' types which the Great Western Railway purchased between 1903–5.

Right: A four-cylinder compound 4-6-2 'Pacific' which found favour on the Nord Railway of France in the early 1900s.

George Whale, who had succeeded Webb on the London and North Western Railway, and John Robinson who joined the Great Central Railway. Both of them turned back to the 4–4–0s which, with the prevailing emphasis now less on speed and more on improved design features, were enjoying a revival. Whale produced his famous 4–4–0 'Precursor' engines, which had inside cylinders and a larger boiler than previous 4–4–0s. Robinson also produced some improved 4–4–0s, which he followed shortly after with his GCR 'Atlantics'. He gained a reputation for producing exceptionally good-looking

Top: The first 'Pacific' to run in Great Britain was the 'Great Bear' produced by the Great Western Railway in 1908. It was the only 'Pacific' used in Great Britain for more than ten years.

Above: A 4-6-0 of the 'Prince of Wales' class designed by George Whale for the London and North Western Railway in 1911. It was considered to be the most generally useful passenger locomotive ever used by this railway.

Right: Following the success of his 2-6-0 'N' class produced during the years of World War I, R E Maunsell went on to develop the 4-6-0 'Lord Nelson' class in the 1920s. They proved an immediate and unqualified success.

LORD NELSON
E
850

engines and the crews who manned his 'Atlantics' christened them 'Lilies' after the lovely Lily Langtry – a favourite of London theatregoers at the time. Some years later, the Great Central Railway reintroduced 4–4–0s, built by Robinson, of the 'Director' class, for passenger and freight work. As with the 4–6–0s he brought out at more or less the same time exclusively for passenger work, they were superheated. The first ten 'Director' class engines were all named after directors of the company. In 1911, Robinson produced some 2–8–0 freight engines for GCR, which were chosen a few years later by the Operating Department of the War Office for military use overseas.

With superheating finding more and more favour among engineers, The London, Brighton and South Coast Railway was running a collection of highly efficient superheated tank engines towards the end of the period between 1900 and 1910. Built by Douglas Earle-Marsh, these 4–4–2s included the '11', '12' and '13' class. When the LB & SCR demanded larger trains to provide a faster service between London and the South Coast towns, a 4–6–2 engine with a longer, larger boiler and larger cylinders was developed from the '13' class.

Earle-Marsh's engines could be described as somewhat experimental and before emulating them,

the London and North Western Railway, for example, invited him to run them on a through trip from Brighton to Rugby. Their performance in maintaining a high average speed, even on difficult parts of the line, while keeping coal consumption low, was impressive enough for the L&NWR to commission the Crewe workshops to build some superheated engines. They built the famous 'George the Fifth' class of 4–4–0s, which in fact were constructed along the lines of the railway's 'Precursor' engines, but differed from them in that they were superheated, and in the design of their cylinders, valves and valve gear. Power output was

An impressive line-up of locomotives of the 4-6-0 'King' class belonging to the Great Western Railway, who introduced them in the late 1920s. They were a larger and more powerful development of the famous 'Castle' class which had gone into service on the GWR a few years earlier and had proved able to cover the first 95 miles out of London in 97 minutes. The GWR had thirty 'King' class engines which regularly worked the West of England and Wolverhampton services, and they were to last right to the end of the steam era, only being withdrawn from the principal express train services in 1958 when general dieselization took over.

KING GEORGE V

sufficiently increased to step up the average speed of trains of the L&NWR from 55 mph to 60 mph with no problem. These engines were quickly followed by the 4–6–0 'Prince of Wales' class engine, which proved to be the railway's most useful passenger-class engine.

These then were some of the engines operating on the British rail network during the first two decades of the twentieth century, and of course there were many, many more classes, built often to deal with the idiosyncrasies of individual lines. The Somerset and Dorset Joint Railway employed some special 2–8–0s which had extra tractive power provided by the very large outside cylinders. The extra power was needed for hauling freight up that railway's 1 in 50 gradients. The North Staffordshire Railway, which had to cope with increasingly large express trains on its steeply graded lines, used an unusual 0–6–4 type of locomotive.

Probably the most immediate and distinctive difference between British and American engines at this time was size – with the British engines as midgets compared to the 'giants' of the United States. The American loading gauge did allow for the building of larger engines (or rather taller, for the track gauge was the same in both countries) but

Above: One of William Stanier's '5XP' 'Jubilee' class of 4-6-0s which was built for the LMS blows out a cloud of black smoke as it pulls up Shap Summit on its way from Blackpool to Perth. Nearly 200 of these engines were built, primarily for intermediate express duties and they were used all over the former Midland Railway system. Each one was given a nostalgic name – the locomotive pictured was called 'North West Frontier'.

Opposite above: 'King George V' – the 'King' class engine that was shipped to the USA in August 1927 to take part in the centenary celebrations of the Baltimore and Ohio Railroad. Construction of this class was speeded up in order that the engine would be ready in time.

Opposite below: This 4-6-0 was photographed in 1937 by which time it belonged to the London, Midland and Scottish Railway. Pre-grouping, however, it had belonged to the Highland Railway where it was known as 'Jones Goods' after its designer – David Jones.

the main reason for the difference was the availability of fuel. The railways of Great Britain had seemingly unlimited supplies of first-class bituminous coal at their disposal, which allowed for high running efficiency from locomotives with only small boilers and fireboxes. As we have already seen, the American railways of this period were more likely to be using inferior grades of anthracite coal, which needed large boilers and huge fireboxes to produce the same amount of steam output.

With the outbreak of World War I in 1914, the British Government took control of the railways and used them with great efficiency to carry troops and ammunition, fuel for warships and other wartime necessities. The scale and speed of operations rivalled those of the American Civil War as trains rattled across the countryside day and night. As might be expected with the whole country involved in war and with the railways playing a leading role, little time was allowed for development work on locomotives. One engineer who did manage to produce a few engines, however, was R. E. Maunsell, of the South Eastern & Chatham Railway. The railway networks through the south-eastern counties were particularly busy and important in conveying men and equipment to the Channel ports. Ironically enough, the SE & CR were using some German 4–4–0s which Maunsell had ordered from Berlin in January 1914. During the war years, he produced his own 'N' class of 2–6–0s to supplement the German engines. Designed for mixed traffic, Maunsell's powerful engines were essentially functional-looking. They were to be the prototype of much new construction after the war, and proved their worth by staying in active service until 1945.

Maunsell was to continue building engines for some years to come. Notable among them were his 'Knights of the Round Table' engines. These two-cylinder 4–6–0s had such romantic names as 'Merlin' and 'Excalibur'. He followed them with his successful 'Lord Nelson' class, which were also 4–6–0s but had four high-pressure cylinders.

At the end of the war, the railway properties were returned to the individual companies but, almost inevitably, were in need of fairly extensive re-organization. An Act was passed by Parliament, which, in 1923 amalgamated all the companies into four groups. The largest of them was the London, Midland and Scottish Railway which comprised the London and North Western, the Midland, the Lancashire and Yorkshire, the Glasgow and South Western, Highlands and Caledonian Railways. The other three groups were the London and North Eastern Railway, which included the Great Central, Great Eastern and Great North of Scotland as well as smaller East Coast companies; The Great Western, which was already in control of most of the west and absorbed the railways of Wales; and the Southern Railway, formed from the London and South Western, London, Brighton and South Coast and the South Eastern and Chatham Railways.

One of Britain's greatest locomotive engineers of all time was now making his presence felt throughout the industry. He was Hubert Nigel Gresley whose love of locomotives had begun when he was a schoolboy and had not faltered since. Gresley actually followed Harry Ivatt as Locomotive Superintendent on the Great Northern in 1911 and for the first ten years of his appointment, he concentrated wholly on producing engines for freight. The war had done much to frustrate his ambitious plans, restricting his capabilities to a narrower field than he would have liked. But the years immediately after the war produced an urgent need for a new fast type of passenger locomotive and Gresley turned his attention to the task. He was both sufficiently young and of sufficiently strong calibre

Right above: Henry Fowler was the designer of this famous class of 4-6-0 locomotives – the 'Royal Scots'. They were introduced onto the London, Midland and Scottish Railway in the late 1920s, and were a great success in spite of the design and construction being hurried to meet the railway's urgent need of more locomotives.

Right: The 'Flying Scotsman' is the most famous of Sir Nigel Gresley's 'Pacifics', and he introduced them to the London and North Eastern Railway shortly after grouping had come into effect. The 'Flying Scotsman' recorded an authenticated speed of 100 mph during one run.

Below: The London, Midland and Scottish Railways first 'Pacific' locomotives were those of the 'Princess' class introduced by William Stanier in 1933. Huge and powerful, they first made it possible for the railway to use just one locomotive for the whole of the 400-mile London to Glasgow run.

MATCHES
CAMP
COFFEE

L N E R
4472

46211

60105
A-3

to ignore the near failure of Churchward's 'Pacific', 'The Great Bear', a few years earlier, and he took the industry by storm by producing some magnificent 4–6–2 'Pacifics' that are recognized as classics of British steam-locomotive history. The first to appear was No 1470, which Gresley christened the 'Great Northern' and it marked the beginning of a new locomotive era in Great Britain. An interesting point about it, however, is that in spite of its immediate impact and the admiration it received from all quarters, it could really be described as a logical extension of earlier locomotives rather than a completely revolutionary design. Based partly on Ivatt's 'Atlantics' and partly on Churchward's 'Pacific', it had a boiler similar to those Gresley had been using in his goods engines. The three-cylinder design, of which he was a great admirer, had seen service before and the high-tensile steel used in the construction of some moving parts, although new to British locomotives, had already been employed in the United States of America.

Some of the fiercest competition Gresley 'Pacifics' were to encounter came from the Great Western's 4–6–0 'Castle' class introduced in 1923 by C. B. Collett. These two types of engine were matched together in exchange trials in 1925 and the smaller 'Castle' proved equal in performance while being economical on fuel. Later Collett followed his 'Castles' with heavier 4–6–0s of the 'King' class. They were rated in fact as the heaviest 4–6–0 passenger expresses ever built in Britain; the first one

Above: In an attempt to keep up with the fashion for streamlining in the late 1930s, the GWR experimented by partially streamlining this 'Castle' class locomotive. This and similar treatment given to a 'King' class locomotive proved quite negative and the shrouds were later removed.

Below: One of the famous streamlined 'A4' high-speed 'Pacifics' designed by Sir Nigel Gresley. They were produced in the Silver Jubilee year of King George V's reign and were given such names as 'Silver Link' and 'Silver Fox'.

Opposite: An earlier example of one of Sir Nigel Gresley's 'Pacifics'. This was one of the 'A3' class – an improved version of his first 'Pacifics' – the 'A1' class.

to be built was the historic 'King George V', which was sent to the United States to take part in the centenary celebrations of that country's pioneer railroad, the Baltimore and Ohio, in 1927. Other competition to Gresley's 'Pacifics' was forthcoming from Vincent Raven's 'Pacifics' run by the North Eastern Railway. They were large and powerful, differing from Gresley's in that they were a natural development of the three-cylinder 'Atlantic' engine. The threat they represented was less serious than that from Collett's engines.

Gresley followed his 'Pacific' No 1470, logically enough, with No 1471, which distinguished itself at a special trial by covering 105 miles in just over two hours, pulling a 610-ton train. The next to appear, No 1472, was called 'Flying Scotsman' and was later renumbered No 4472. With the grouping of the railway companies now well enforced, Gresley was of course attached to, and therefore producing his engines for, the London and North Eastern Railway. Soon the old rivalry of a couple of decades earlier came gently to the fore and the LNER began to experience some competition from the London, Midland and Scottish railway on the route between London and Scotland. The engines mainly used by the LMS towards the end of the 1920s were Henry Fowler's 4–6–0 'Royal Scot' class, which this engineer had produced in cooperation with the North British Locomotive Co. Simple in design, they were nevertheless a great success and included such historic locomotives as the 'Lancashire Witch'.

Having largely survived the slump of the 1920s, the railways had to face up to the fact that other forms of transport were beginning to present serious competition – in particular long distance motor coaches, whose fares were highly competitive with those of the railway. Gresley made a trip to Germany to study the new diesel engine that was running on the Berlin–Hamburg line and which he thought might be the answer in restoring the superiority of the railways. Impressive though its performance was, Gresley returned to Britain unconvinced that it was better than anything that could be produced by steam locomotion. However, he had been told to make improvements in the LNER's service to the north, both in terms of time and passenger comfort, and he began to devote his attention to this task. In 1934 he produced one of the most powerful passenger locomotives ever built in Great Britain – a 2–8–2 wheel arrangement called 'Cock o' the North'. On tests this engine pulled 160 tons, touching a speed of 85 mph at one point on the line. This however, was beaten by the 'Flying Scotsman', which touched 100 mph pulling a special four-coach train weighing over 200 tons during a record run between London and Leeds. The whole journey took just over 2½ hours. In 1935, an improved 'Pacific', Gresley's No 2750, called 'Papyrus', proved that steam power was still very

Above: Under construction – William Stanier's streamlined 'Pacific' 'Coronation', produced in 1937 for the London, Midland and Scottish Railway, is seen here on the factory floor at the Crewe Works.

Right: Complete and resplendant in Prussian blue livery with silver white streaks, Stanier's 'Coronation' is admired by onlookers. She was produced in the Coronation years of King George VI, and was later re-painted in the traditional red of the LMS in order to take part in an American exhibition.

CORONATION

much a going concern by working a round trip from King's Cross to Newcastle in one day, thereby cutting an hour from the existing best time for the journey. During the run it reached a speed of 108 mph which was quite an achievement.

Gresley was still not wholly satisfied that he could introduce and maintain a top service with these engines, and began to turn his attention to streamlining. Streamlining had already been acknowledged as a means of reducing air resistance encountered at higher speeds, and although it was in its infancy, its appearance found favour as it fitted in with the fashionable ideas of the time. It was therefore felt that if it could be introduced into locomotive design, it would do much to help the image and prestige of the railways. The resulting publicity might even convince the growing band of cynics that the industry was, after all, keeping up with the threatening menace of the car and the aircraft. Some attempt at streamlining had been made with Collet's 'Castle' and 'King' classes for the GWR series – streamlined casings had been applied to some of these engines in an attempt to reduce wind resistance. Gresley, however, modified his earlier 'Pacifics', incorporating various improvements to the boiler, firebox and cylinders and then adding the streamlined casing. The result was the 'A4' class of which the first of the initial four engines was 'Silver Link'. The engine was described as being 'whale-shaped, with its front end resembling a gigantic wedge'. This caused a strong up-current of displaced air when the engine was in motion. The engine was painted a silvery grey and the seven coaches of its train were all painted to match. Ready in the year that marked King George V's jubilee, the whole train was called the 'Silver Jubilee' and was to go into regular service at the end of September. Three days later, however, it was scheduled to run on a special trip, which it did, watched by thousands of spectators. During the course of it 'Silver Jubilee' set up a new rail speed record of 112½ mph.

The racing spirit among railway operators had been awakened yet again and the LMSR topped 'Silver Link' by recording a speed of 114 mph near Crewe in 1937. In doing so, however, they had only established a new United Kingdom record, for a 4–6–4 in Germany had recorded a speed of 118 mph in 1936. In his calm way, Gresley rose to the tacit speed challenge and in 1938 another of his 'A4' streamlined 'Pacifics', the 'Mallard', set out on an apparently routine run to test a new form of vacuum brake. In fact, the crew were going out in an attempt to 'crack' the record, and crack it they did with a speed for a steam locomotive that remains unsurpassed today: 126 mph.

The first four of Gresley's 'A4' Pacifics had all been the distinctive silver-grey colour, but the next series to be built, of which the 'Mallard' was one, were a 'Garter' blue colour. This became the standard colour for the class from then on.

For the coronation of King George VI in 1937, the LNER produced another high-speed train called, aptly enough, the 'Coronation'. In the same

year, its old rival the LMSR answered with its version of the streamlined 'Pacific', designed by William Stanier who had succeeded Henry Fowler. They were extremely sophisticated engines, painted in blue and silver, and it was one of these in fact, No 6220 also called 'Coronation', that recorded the 114 mph on a test run for LMSR.

The A4 'Pacifics' were not the last engines Gresley built. He followed them with his beautiful-looking engines, called the 'Green Arrows', which were built in various forms for mixed traffic use. His last engine was a 2–6–2 wheel arrangement known as the 'Bantam Cock'. He died in 1941 having had many honours, including a knighthood, bestowed on him, and while he was actually at work on the design of an armoured carriage for Winston Churchill's use.

The outbreak of World War II in 1939 more or less brought an end to streamlining, which proved to have little functional wartime value. Most of the casings on the streamlined Pacifics in existence were removed either wholly or partially and the bright colours replaced by a more fitting, but austere, black. O. V. S. Bulleid, who had replaced Maunsell on the Southern Railway, was one engineer who continued to build engines during the war, notably the 'Merchant Navy' class 'Pacifics' which did have some streamlining initially. Bulleid's appointment had come at a time when the Southern was extending its electrification. He was a devotee of steam, however, and he managed to effect something of a revival for this form of power to the Southern Railway. His 'Merchant Navy' engines were fast, powerful 'Pacifics' designed to use low-grade coal. After the war, lighter versions of the 'Merchant Navy' were built, including the commemorative 'Battle of Britain' locomotive.

The golden days of steam locomotion, brought to the peak of its glory and glamour by the high-speed streamline era, were now over. By the end of World War II the very days of the steam engine were numbered. When peace came, services were restored, and the 'Liberations' – austere-looking 2–8–0s built during the war for the Ministry of Supply – formed the main bulk of steam engines in use. In 1948, following the nationalization of the railways, all the companies came together under the general heading of British Railways.

The first engine designed by the engineers of British Railways was the last steam engine to go into regular service. It was a class of 'Pacific'-type, called the 'Britannia' (to emphasize the 'national' rather than individual aspect of it) and it embodied all the features that had proved to be the most satisfactory from former engines. Successful though

Streamlining locomotives became fashionable in the 1930s, but brought with it a variety of problems. Not the least of these was the possible effect upon the engine workings brought about by the elimination of the natural flow of air normally encountered at speed. During World War II, the casings were removed entirely from the LMS engines and partially from those of the LNER. In the foreground is a streamlined 'Pacific' built for the Pennsylvania Railroad.

70023
BACON FACTORY

92029

these engines were, their career was cut short by the decision in 1955 to supersede steam traction with diesel and electric. Steam wasn't quite beaten, however, for the final steam-powered locomotive to be built was completed after that decision. It was a 2–10–0 design produced by the Swindon works and the last one ever made was No 92220, called 'Evening Star'. It took to the tracks of the Great Western 120 years after the first engine – 'Morning Star' – had inaugurated travel by steam locomotion on this great, and by then, historic, line.

Opposite above: The first new design of the nationalized British Railways – a 'Pacific' known as the 'Britannia' class. They were introduced only a few years before the decision was made to phase out steam.

Opposite below: The last steam locomotive to be built in Great Britain belonged to the 'BR9' class of 2-10-0s.

Below: One of the 'BR6' Pacific 'Clan' class built by British Rail to fulfil a need for a less powerful, lighter locomotive than the 'Britannia' class.

Overleaf: One of the 'Duchess' class of Pacifics, developed from the 'Princess' class and built for the LMS in the late 1930s.

CITY OF CARLISLE
46238

46238

THE AUSTRALASIAN STORY

Below: The 'C32' class of 4-6-0s of which three are seen here running an excursion trip in Australia's New South Wales were the stalwarts of passenger working engines in this State for more than thirty years. They originated as the 'P' class, built in the 1890s by a GB firm and the design was so successful that companies in Australia and the USA began to build them, with continual improvements over the years. In the 1930s, some were assigned to special duties and repainted in bright colours, thus breaking from the previous tradition of black livery.

Right: A 2-6-6-0 Mallet steaming its way over the mountains of Java, Indonesia. It was built by the Dutch railroad and was used to pull mixed trains.

Overleaf: The 4-8-4 wheel arrangement was a favourite in New Zealand. Here a 4-8-4 of the 'KB' class crosses the Waimakariri River.

Above: The electric wire overhead bears testimony to a new age of locomotive power, but a 'J' class 4-8-2 steam engine in Otira, New Zealand, continues operations.

Right: A fully streamlined version of the '38' class Pacific seen below left.

Below: The Beyer-Garrat steam locomotives found much popularity with the railways of the New South Wales Government. Here one is seen on the Orange-Dubbo line.

Left above: One of the semi-streamlined 'Pacifics' of the '38' class owned by the New South Wales Government Railways. They were used for both passenger and freight services.

Left below: A 'Pacific' of the New South Wales Government Railways' '38' class painted in the alternative colours of green with yellow trim.

Above: On the way to Arthur's Pass a KB-class 4-8-4 crosses the Wainakarin River.

Below: Mallet engines originated in France and were first used on European railways at the very end of the nineteenth century. Shortly before World War I, a Bavarian firm began to manufacture them with various wheel arrangements and some of these were exported to Indonesia – then the Dutch East Indies. Mallet engines proved ideal for the railways of Java and when larger locomotives were needed, they imported more Mallet types from Holland and the USA – in particular the 2-8-8-0 wheel arrangement type, pictured here on a run from Bandung to Tijibatu.

Right: A 2-6-6-0 Mallet engine built by the Dutch Railroad for the mountainous railways of Java.

CC 50 17

CC 5017
CC 5017
C 2724

Left: An oil-burning 2-6-6-0 leads a mixed train near Nagrak, Java.

Left below: A 4-6-4 Tank engine on the narrow tracks of an Indonesian Railway.

Top: A relic of the past – an old 4-4-0 steam locomotive that has done long service in Java.

Above: More German-built locomotives in Indonesia. This one, a 2-8-2, has smoke deflectors included in the design. They were intended to create an airflow to lift the smoke above the cab.

Although steam-locomotive travel on railways began some quarter of a century later in the great land mass of Asia than in Britain, it would be inaccurate to say it had no significant part to play in the history of steam locomotion or, for that matter, railways in general. The diverse problems of early rail travel initially encountered throughout Asia were met with the same pioneering determination and tenacity that had enabled railway engineers to develop a viable network in the small overcrowded island of Great Britain, and open up the vast, inhospitable west of the United States of America.

Let us begin in this area with the world's smallest continent – or largest island – Australia. At a time when steam locomotion on railways was a recognized means of travel in Britain, Australia's first white settlement had been in existence for less than fifty years. Most of the country was still undeveloped and uninhabited, as indeed a great percentage of it is today. Yet the linking of far-flung towns and remote areas and the opening up of farming or mining districts depended upon the construction of some sort of rail system. Australia's first railway was

C62 2

Left: Cold weather conditions present no problem to the Japanese 'C62' 4-6-4s, double-headed to pull an express train. The 'C62s' were built in 1948 and 1949, but incorporated boilers from some of the wartime assigned 'D52s'. The 'C62s' were considered to be among Japan's finest and largest locomotives and were not withdrawn from service until 1971.

Right: More than a thousand of these 'D51' 2-8-2s went into service on Japanese National Railways between 1936 and 1945, to become one of the most common steam locomotives ever used in this country. They presented a logical progression from the smaller 'D50' 2-8-2s, and many were rebuilt after World War II to improve on their utilitarian war-time lines. This one is pictured at Oshamambe.

Below: An attractive setting for this 'D52' 'Mikado' as it steams round the coast off one of Japan's islands. The 'D52' was one of the last steam locomotives to be withdrawn from service in Japan.

C 57 104

in the State of Victoria – a $2\frac{1}{2}$-mile line from Port Melbourne to Flinders Street, opening in 1854 and constructed to a 5 foot 3 inch gauge. It was followed a year later by a 14-mile line from Sydney to Parramatta in New South Wales with the standard gauge of 4 feet $8\frac{1}{2}$ inches. Thus the diversity of gauge had begun already, and the vagaries were destined to get worse as construction progressed!

Above: One of the first of four engines imported into New South Wales, Australia from Great Britain in 1855. They were built by Robert Stephenson & Co and were similar to the 0-6-0 type he had produced for the Stockton and Darlington Railway. Although the engines would be required to travel over greater distances in Australia, the working conditions were in fact considered to be similar to those on the Stockton and Darlington Railway.

Left: A 'C57' climbs Karakachi Pass with a passenger train; a 'D51' helper on the rear is buried in the smoke.

Initially, as railways were planned in Australia, a recommendation from the governing body in the home country had suggested that the standard gauge of 4 feet $8\frac{1}{2}$ inches should be used throughout Australia. New South Wales and Victoria subsequently both decided to adopt 5 feet 3 inches and the London powers-that-were gave their consent. Planning of the lines got underway and though they were hundreds of miles apart, there would be no future problem in linking the two States by rail. Until, that is, the manager of the Sydney line, an ardent Scot, persuaded his team to revert to standard gauge.

The States of Australia developed autonomously, and the individual governments, conscious of the ever-present inter-state rivalry as well as the urgent need to provide a railway to allow their own areas to be opened up and developed, built their tracks to whatever width suited them best at the time. Little thought was given to the day in the future when links between States would make unification desirable. In consequence, Victoria went on to build over 4,000 miles on their 5 feet 3 inch gauge, while New South Wales laid 6,000 miles of standard-gauge track. Queensland and Western Australia, more sparsely settled as they were, decided they could not afford either gauge and instead laid tracks of 3 feet 6 inches. South Australia possessed the third railway to be built in Australia, from Port Adelaide to Adelaide itself, which was constructed to a

Above: The 4-4-0s first ran on the railways of New South Wales in the late 1870s, the original ones being designed by the British firm of Beyer, Peacock and Co. as a development of their class of 2-4-0 passenger engines built in mid 1860. The 4-4-0 pictured is pulling a first and second class compartment as well as a sleeping car and brake van.

Left: The Beyer-Garrat 'articulated' locomotives were amongst the largest locomotives ever used in Australia. This one is hauling a string of coal-laden trucks near Dubbo, out in the west of New South Wales.

5 foot 3 inch gauge. South Australia also constructed railways of the 3 foot 6 inch gauge, and when, some while later, the Commonwealth decided to build a rail link between the east and west, the 4 foot $8\frac{1}{2}$ inch gauge was used. Port Pirie in South Australia thus became a unique junction, the meeting point of the three gauges!

Railway building in Australia proceeded rather slowly in comparison with other parts of the world and even today a large percentage of the lines built is coastal. The first rail link between two capital cities came nearly thirty years after the opening of the first railway. It was between Melbourne and Sydney in 1883, and was followed four and six years later by links between Melbourne and Adelaide, and Brisbane and Sydney respectively. The notion of a trans-Australian railway was conceived in 1870 and the major part of it – over 1,000 miles between Port Pirie in South Australia and Kalgoorlie in Western Australia – was built in a 5-year period. The fact that no tunnels or bridges were necessary greatly facilitated the building and the line incorporates the longest, straight stretch of track in the world – more than 300 miles without a curve. The final links to Perth in the extreme west and Adelaide in South Australia were completed and the line opened for traffic in 1917.

In addition to the east-west link, a north-south line was also planned but has never reached ultimate completion. The line extends 800 miles or so north from Adelaide to Alice Springs in the centre (with at least one change of gauge still to be negotiated by travellers in this stretch) and approximately 300 miles south from Darwin in the extreme north to Birdum, where it stops seemingly in the middle of nowhere. Traffic between Alice Springs and Darwin therefore has to go by road for at least 500 miles of the journey.

Australia's first steam locomotive ran on the Sydney–Parramatta line in 1855. Imported from Britain and considered suitable for mixed traffic, it was built by Robert Stephenson, who based it on his six-wheeled 'Patentee' design of the 1830s. In the following ten years, New South Wales Government Railways had ordered several 'long-boilered' 0–6–0 types from Robert Stephenson and Co. They were

A 'C36' engine pulling the Newcastle to Sydney express. This class was introduced into Australia in 1925 and became one of the most successful passenger express locomotives ever to work in the country. They had runs of up to 100 mph incorporated into their regular schedules and worked all the principal expresses until the 'C38' class was introduced in 1948. They were also chosen to service the Commonwealth line linking South and Western Australia across the Nullabor desert when a powerful locomotive was needed in 1937.

3664

needed to meet the demand for powerful goods engines, which was brought about by the severe grading in places along the line as it was extended south from Sydney. To describe these engines as hard-working and long-lived must be one of the greatest railway understatements, for at least one of them continued in active service until it was 93 years old!

The passenger-service engine contemporary to Stephenson's 0–6–0 goods engine was the 'G' class 2–4–0, first ordered in 1865 from the English firm of Beyer, Peacock & Co, which was to be the main overseas supplier of locomotives to Australia. Similar to 2–4–0 tank engines in use in Great Britain at the time, the Australian version was additionally fitted with a large canopy over the footplate to provide shade from the relentless sun. A few years later, the NSWGR ordered some '79' class 4–4–0 locomotives from Beyer, Peacock & Co. These engines were a direct development from the 'G' class, but were fitted with a larger four-wheeled leading truck, instead of the two-wheeled 'pony' truck.

One of the most successful engines built by Beyer Peacock for the NSWGR was the 'C32' class of 4–6–0s, first introduced towards the end of the nineteenth century as the 'P' class. They followed an assignment of 2–6–0s, class '205', also built by Beyer Peacock, which did sterling work as goods engines. Some years later the 'C32' design, which received constant improvements during its long service, led on to the building of the 'C35' 4–6–0s. Slightly more powerful engines, they were initially brought into service to cope with the effects of the heavy coaches and dining and sleeping cars that passenger locomotives were having to haul on their ordinary services.

Great Britain wasn't the only country supplying Australia with locomotives. The prevalent feeling among Australian railway men that their operations were more similar to those in the United States than the home country led to orders being placed with America's Baldwin company. Some of the first engines supplied by Baldwin were 4–4–0s, which represented competition to Beyer Peacock's '79' class. These were followed by some 4–6–0s, and subsequently by powerful 2–8–0s in about 1879. There was never any difficulty in spotting an engine's country of origin for each retained their inherent characteristics, even though they were designed for Australian conditions to specifications laid down by Australian engineers.

The nature of the service required from the railways differed from state to state and the locomotives differed accordingly. Victoria, faced with an increasingly busy suburban service around the capital of Melbourne, met its needs with efficient little 2–4–2 tank engines. The 1860s also saw

Victorian Railways using the 'B' class 2–4–0 engines, often known as 'Old English'. Some were built by Beyer Peacock and although extremely ornate in appearance, they were equal to the long, hard service that was demanded of them. They were largely replaced by the 'Old A' class of 4–4–0s which were introduced in the 1880s, and were similar to the NSWGR's '79' class of 4–4–0s, whose fame had spread southwards. Beyer Peacock built a consignment of these engines to suit Victoria's wider gauge, and as high speed passenger locomotives, they enjoyed the success they had found in the adjoining state. They were sometimes used in conjunction with their predecessors, the 2–4–0s, to haul very heavy trains.

Above top: A Queensland Government Railways 'C17' 4-8-0 on the turntable at Ipswich. Although introduced in 1920, this locomotive, designed for the State's narrow gauge, bore characteristics reminiscent of locomotives of a much earlier period. Nevertheless new assignments of the basic design were delivered up until the 1950s and they were used as standard freight engines into the 1960s.

Above: A 'C17' pulling ventilated poultry cars through Queensland.

Left: Imported from Germany, the 'J' class 2-8-0 was used by Victorian Railways for freight service, particularly to transport the State's wealth of wheat. They were only withdrawn from service at the end of the 1960s.

Although progress into the twentieth century did not bring the end of American and British engine import, it did mark the beginning of Australian locomotive manufacture. One of the first to be produced was the 'A2' 4–6–0 type, built at Newport Works in 1907 for Victorian Railways. It was a totally Australian enterprise and proved so successful that over a hundred more were built between 1908 and 1915. It led also to the building at the end of World War I of a companion locomotive designed for heavy freight service. A 2–8–0 wheel arrangement, designated 'C' class at the time of going into service, these were the heaviest and most powerful engines in Australia and looked very distinguished with their handsome scarlet paintwork. They led on to the goods locomotive class 'X' which were 2–8–2s. Their austere lines replaced the more decorative and attractive earlier designs and the addition of the gaunt smoke-deflecting plates placed either side of the front of the boiler, currently in favour in Britain and continental Europe, did little to enhance the locomotive's appearance. Some of the gracefulness of previous engines returned in the design of the 4–6–2 class 'S' passenger locomotive, which was also introduced at the end of the 1920s.

Twelve years later, Victoria Railways produced the huge 4–8–4 class 'H'. Nicknamed 'Heavy Harrys', they proved too big for most lines and there was no money in the railways' kitty to give the tracks the strengthening needed to support them. The engines instead were replaced in the early 1950s by a British-built engine – the 4–6–4 class 'R' which in spite of its Scottish ancestry, looked very much more like engines of contemporary German design.

The NSWGR also began to build its own locomotives and in mid-1920s it produced one of Australia's most famous express classes. It was the 'C36' 4–6–0, ten of which were built by the railway's works and sixty-five by an Australian

company, Clyde Engineering. In their original form, they were virtually unrivalled in express-passenger service for twenty years, until the introduction of a 'C38' class. However, they were later rebuilt to incorporate the Belpair firebox and in this form were extremely powerful engines. Their short, blunt shape earned them the nickname 'Pigs'. At the end of the 1920s, NSWGR put some very powerful 4–8–2s built by Clyde Engineering into service for heavy freight work, and the 1930s saw the introduction of the three-cylinder engine design and the use of mechanical stokers to satiate the hungry fireboxes.

The 'C38' classes mentioned above did not come into service on the NSW network until the end of World War II. A 4–6–2 'Pacific'-type engine, it was designed with a streamlined casing, as of course were engines of this era in other parts of the world. It had a bullet-shaped nose cone and a cowling over the top of its smooth boiler jacker, as well as a thick skirting over the running boards. Later designs did not have the cowlings or the bullet noses.

During the course of their history, at least two engine types doing service on the tracks of NSW were chosen by the Commonwealth Railways who operated the great transcontinental line that linked South and Western Australia. The first ones were needed when the line was opened between Port Augusta and Kalgoorlie in 1917, and 2–8–0s of standard NSW design were ordered from the North British Locomotive Company for freight traffic. The 'C36' was chosen in 1937 when a powerful locomotive was needed for the passenger expresses.

An extremely interesting locomotive was designed and put into use in Australia at the beginning of the twentieth century. Known as the 'Garratt', after the Australian inventor who patented the design in 1908, its single large boiler was suspended in a frame supported at either end by two separate engines to which driving wheels were attached. One engine unit incorporated the fuel tank and the other the water tank while both were served by one steam source. In effect it was an alternative to the Mallet articulated engine, for it was also articulated and could thus negotiate sharp curves. It was ideal for use on light rails, as the weight was distributed over a large area and therefore did not rest directly on top of the driving wheels. The Garratt design was first incorporated into an 0–4–0 + 0–4–0 type built by Beyer Peacock for use in Tasmania, the small island to the south of Australia's mainland, which had 2-foot gauge tracks. Beyer Peacock later built some 4–4–2 + 2–4–4 types for the Tasmanian Government Railways and later still, in the 1950s, the company produced the 4–8–4 + 4–8–4 'AD 60s' for NSWGR. These were the largest, although not the most powerful, Garratt engines ever built and they were used on the route north from Sydney to Newcastle which had two extremely steep gradients. These the Garratts took with barely a change in pace, just as they happily hauled heavy coal and ore trains on the Orange to Dubbo line to the west of Sydney.

The early years of the twentieth century found a

These unusual looking locomotives – South Australian Railways 520 class of 4-8-4s – were built at the Islington Works, SA and first appeared in 1944. Although they look large and heavy, they were in fact only moderate-sized locomotives, weighing just over 100 tons without the tender. Had the boiler been as long as it appears, the heating surface would have been much greater than required and the engine would have been correspondingly too heavy for the track it was to operate. They were used on South Australia's railways for just over twenty years.

very unforthcoming locomotive policy in South Australia, where the low axle loading permitted had frustrated and restricted engineers for some time. The company received its long awaited injection of vitality in the form of an American, W. A. Webb, who took over as Commissioner of Railways in 1922. The railway's need was for large engines to cope with the heavy trains crossing the Mount Lofty Range and its engineer designed two huge new machines of 2–8–2 and 4–8–2 wheel arrangements. They looked characteristically American in appearance, yet they were built in Britain to wholly Australian specifications. They did long service, and almost the next new engine the railway was to see came some twenty years later in the form of the striking-looking 4–8–4s, which were streamlined with a yacht-like prow over the smokebox. An interesting fact about South Australian Railways is that although by no means the most advanced of Australia's railways, it was the first large system to begin using diesel-powered locomotives.

The two remaining main railways in Australia, those of Queensland and Western Australia both had narrow 3-foot gauge tracks. Some of the Western Australia's best-known locomotives were the 4–6–2 'Pacific' Class 'P's imported from Great Britain in the mid-1920s. The design was copied by the railway's own workshop and the engines did good service for ten to fourteen years. They were then redesigned to provide a more powerful engine and renamed Class 'PR'. In the 1940s, Western Australia needed powerful locomotives to haul heavy coal traffic and the railway ordered some large 4–8–2s.

Queensland's main use of her railway was to serve the cattle country. The two engine types most extensively used were the 4–8–0 'C17' class introduced in 1920, and the 4–6–2 'B18' class brought in some six years later. The 'C17' was more reminiscent of locomotives designed at the turn of the century than those more generally recognized as designs of the 1920s, but this only served to emphasize the nature of railway operations in Queensland. Not until the coming of dieselization and the construction of the new, heavy-duty lines needed to carry the mined wealth of the State did Queensland's railways take a giant leap forward.

Even in the modern diesel and electric age, when steam in Australia has met the same fate as in Britain and America, the railways of Australia still have to contend with the old gauge discrepancies. Standardization is on its way, but with a massive £280 million estimated as long ago as 1950 for complete conversion, it is not hard to understand why it may take a little time to arrive.

New Zealand with its rugged mountains, great rivers, deep-cut valleys and terrain that rises rapidly from the coast has presented railway engineers with

many headaches. These can perhaps be epitomized by the severe grades of the Rauriumu Spiral in North Island and the famed Arthur's Pass in South Island. Lines of various gauges have been built but a standard of 3 feet 6 inches was adopted in 1860, in part at least because it was the most economical way of tackling the mountainous districts. The first line was opened in New Zealand in 1863, and seven years later the country could still boast only 46 miles of track. In the next eighty years, however, the figure was to increase to over 3,500 miles, of which none was ever more than 75 miles from the coast.

The major role of New Zealand's railway has

Left: The 'Pacific' type of locomotive was first used on the railways of New Zealand early in the twentieth century, and the same design with various modifications and improvements were used into the second half of the century. In fact scrapping of them did not begin until 1956. In their hey-day they handled much of the main-line passenger traffic as well as most of the goods services. The first examples were built at the New Zealand Government Railway's workshop at Addington and the illustration shows one of the later modified forms, although the changes made to the basic design throughout its long life were minimal.

Below: The New Zealand Government Railways' 'K' class of 4-8-4 was brought out in 1932 when it became apparent that a larger locomotive than the successful 'Pacific' was needed for heavy-duty service on North Island. The specifications for these engines stated they were to be fifty per cent more powerful than the 'Pacifics'. Thirty of them were built and put into service in the four years from 1932 to 1936.

probably always been the transport of goods. Passenger traffic is important between the largest cities, but in suburban areas and over short distances, buses and cars have been more generally used since their inception. New Zealand railways, however, have seen a variety of locomotives, of which some early ones were the 0–6–0 tank engines, built in Leeds, England. Originally intended for work in the southernmost part of South Island, their reliable service led to their being transferred further north to Christchurch (by sea!) for passenger and goods use. They were later rebuilt to a 2–4–4 wheel arrangement which made them more suitable for work on lighter tracks. Also imported from Britain were the powerful 'P' class 2–8–0s, used for freight work, many of which did over forty years of service.

New Zealand did not just import from Britain, however. As in Australia, some engineers felt working conditions were more akin to operations in the United States and accordingly, in spite of patriotic opposition, engines were ordered from American companies. Notable among these were the 'K' class 2–4–2s, imported in 1877, the last of which ran until 1928, and the 'N' class 2–6–2s imported in 1885. These more powerful engines were built by the Baldwin company and were used for both passenger and freight work.

Although the railways of New Zealand used imported trains for the first twenty-five years or so of operations, in 1889, New Zealand Government Railways, which more or less controlled railway operations nationally, built its first locomotives. They were tough little tank engines with a 2–6–2 wheel arrangement, especially designed to provide banking assistance on some of the very steep inclines. In the early years of the twentieth century, the same workshops produced the famous 'Ab' class of 4–6–2 'Pacifics', the outstanding design of which enabled them to run in virtually their original form for over forty years. Their success led to the construction of a 4–6–4 tank engine version, which was used for freight and suburban passenger work around North Island's Auckland and Wellington. When it became apparent towards the end of the 1920s that larger locomotives than the 'Ab' Pacifics were needed in North Island, the massive 4–8–4 'K' class emerged. It was also developed into the 'Kb' class by mounting a small, separate steam engine, or a booster, in the trailing truck. This was connected to the trailing wheels, which could therefore supply additional tractive force – an extremely useful adjunct for combating steep grades. The 'Kb' class was used in South Island for operating the undulating grades on the Christchurch side of Arthur's Pass.

The 'K' class engines were followed by the streamlined 'J' class of 4–8–4s, which had a lighter axle loading and were therefore more suitable for tracks with lighter rails. The streamlined case was more to keep up with world trends than actually to serve the cause of speed, for the engines were rarely to exceed 50 mph.

Steam locomotive power retired later from New Zealand than in many parts of the world. In 1960, when few lines in America or Britain were still using steam engines, it still predominated in New Zealand. Not until the early years of the 1970s did steam finally give way to modern diesel and electric engines on New Zealand's main lines.

Steam locomotion as a motive power doesn't only belong to a bygone era. A journey of a thousand miles or so northwest from Australia finds a land where steam engines are still doing useful service – in Indonesia. Railways came to the Indonesian archipelago about a hundred years ago when it was the Netherlands East Indies. Initially the Dutch opened up the flat coastal plains of their colonial empire, laying tracks of 3 foot 6 inch gauge. With no steep gradients to combat, small lightweight locomotives were quite adequate. As the railways began to penetrate the mountains, more powerful engines were needed to climb the steep gradients, particularly prevalent in the west. In the early years

Above: One of the smaller Mallet locomotives to run on the railways of Indonesia. It is an 0-4-4-2T locomotive of the type imported from Germany in the early 1900s. Mallet locomotives in Indonesia have the distinction of being among the last of their type to have operated anywhere in the world.

Right: A coal train leaving the Singareni coal fields, India.

of the twentieth century, 0–4–4–2 Mallet articulated engines were imported from Germany and these were soon supplemented by larger tank engines from Holland. When still larger engines were needed to service the increasing demands of the network, 2–8–8–0 Mallet articulateds were imported from America, built especially for the 3 foot 6 inch gauge. The numerous sharp curves of the tracks made articulated engines particularly suitable and after World War I still more 2–8–8–0 and the lighter 2–6–6–0s were imported from Holland, Germany and Switzerland.

In an area rich in oil wells, it was not surprising that the Indonesian railway men made a fairly early decision to change steam for diesel – shortly after the country had gained independence in fact. Plans

were thwarted, however, under the rule of President Sukarno which even resulted in a deterioration of locomotives and railways. Although today the rail system in Java particularly is remarkably well-developed, with no area as much as 50 miles from a railway line, many of the old steam locomotives are still in service, becoming the last of their kind to be so anywhere in the world.

Travelling still further west we find a country with many claims to railway history fame – India. Two of the most notable claims might be the fact that their railway system is exceeded in size only by those of North America and the USSR, and that India was the first, and possibly will be the last, Asian country to use steam locomotive power. Railway construction began in India in the 1850s, in spite of a feeling both in India and in the British government that it was the waterways that should be developed for the future transport system. In 1853, however, the first line from Bombay to Thana – destined to become part of the Great India Peninsular Railway – was opened, soon followed by lines around Calcutta, Madras and Lahore. The original recommendation was for a standard gauge of 5 feet 6 inches, but various factors led to a metre gauge and even narrower – 2 feet 6 inches – being adopted in some areas. This was less because lines were built independently, as in Australia, for example, than as a result of strategic planning and considerations. It was often felt that the cost of construction over difficult terrain in areas where the use of the service was likely to be minimal did not justify the laying of broader-gauge tracks. Although the mixture of gauges brings the inevitable protracted changeovers in train journeys, it is not as inconvenient in India as it may appear. All the most important cities, ports and main producing areas are linked by broad-gauge lines, albeit often only single track.

As in many countries the first ten years of operations saw railways constructed at a fairly slow pace. By 1863, just 2,500 miles of track was laid. At the time of partition in 1947 this had grown to over 40,000, of which some 20,000 was of broad gauge, 16,000 of metre and 4,000 of narrow gauge.

For many years, understandably enough, all railways were serviced by British locomotives, of which the first to run on the Bombay to Thana line were 2–4–0s of the simplest style of construction, supplied by the Vulcan Foundry Ltd. The East Indian railway, which emanated from Calcutta, used 2–2–2 express locomotives to handle its most important service, the mail trains. The North Western Railway, which became state-owned when it purchased the Scinde, Punjab and Delhi Railway and amalgamated it with three smaller lines, favoured 0–4–2s for mixed traffic requirements. In the latter part of the nineteenth century, most broad-gauge lines were using 4–4–0s or 2–4–0s for passenger and mail services and 0–6–0s for goods work. Then 0–6–0s began to be introduced to cope with the increasing heavy passenger traffic and they led to the development of the 'A' class 4–6–0s which were found ideal for mixed traffic.

As may be imagined, with the railways the economic backbone of India throughout the twentieth century, the range of locomotives used has been quite as diverse and varied as in other parts of the world. They have included various expressions of the universally popular 4–4–0s and 4–6–0s for passenger service, tank engines of various wheel arrangements, as well as 2–8–0 + 0–8–2 and 4–8–0 + 0–8–4 Beyer-Garratts; the last were the largest and heaviest locomotive ever to run on Indian railways. In the 1920s the first Indian

Above: A 'D51' Mikado – one-time Japan's most widely used 'all-purpose' locomotive – is seen here pulling a freight train through the snowy scenery off the northern island of Hokkaido.

Right above: A 'C62' class 4-6-4 at Sopporo in Japan. Built to pull the country's express trains, they found popularity in many other countries too.

Right below: A 'C61' class 4-6-4 at Taira, north of Tokyo – one of the last outposts of steam operation in Japan. It was here that they handed over to electric-engines which completed the run into Tokyo.

Locomotive Standard Committee was set up, appointed to produce standard designs for use on any Indian railways, although not all were Government-owned at the time. The committee recommended the use of 'Pacific' type locomotives, which became the standard passenger engine throughout India. Built in many different forms since then, the latest ones are characterized by a bullet nose and a streamlined casing.

From India we may go to the islands of Japan to find another interesting story in the history of the railways. When the feudal rule of the Samurai was replaced by an enlightened government in the 1870s, help was sought from Britain to construct a railway line from Tokyo to Yokohama. In 1872 the seventeen-mile line was opened and serviced by neat little 2–4–0 tank engines known as 'Vulcans'. For the next twenty years the building of railways was slow, with a sudden boom period in 1894–8. Today Japan has more than 22,000 miles of track and the train passenger can travel through the four islands from north to south without ever leaving his compartment. In addition Japan can boast the fastest point-to-point schedule in the world from Osaka to Okayama as well as the world's most crowded rail system – overtly demonstrated by the fact that professional 'pushers' are employed on the Tokyo services to cram passengers into the compartments before the doors can be closed! But for the steam enthusiast, as in Indonesia, steam has lived on longer than in the West, this time to do battle with the winter snows in the mountains between the seaport of Hakodate and Hokkaido's capital of Sapporo.

In 1897, British-built 4–6–0 engines were in use on the Tokyo to Yokohama line, and these were joined by the famous 2–8–2 'Mikados' built by Baldwin of the United States (see Chapter 2). In 1906, the variously owned railways of Japan, including the Imperial Government system, amalgamated into the Japanese National Railway. Various forms of 4–4–0 locomotives were among the most popular express passenger engines – some emanating from Britain and others from America. Several years later they gave way to the home-built massive-looking 'C62' class 4–6–4s, which for a long time had colourful headboards bearing the emblems or name of the train.

The famous 'Mikados' were built in Japan in the 1930s and the 'D51' class became the most common steam locomotive, produced as it was with many different modifications. Some even had semi-streamlined casings, although this was generally done away with during war years. Today, however, in an era that feels centuries away from steam locomotion, Japan appears to be leading the railway world with her famous electric 'bullet' train which averages 112 mph on the Tokaido line.

STEAM LIVES

Great Britain

It did not need a clairvoyant to forecast in the early 1950s that the era of railway steam locomotion in GB was progressing towards the realms of history. In fact the giving-way of steam power to the more modern forms of diesel and electric traction was a gradual process which had already been effected over parts of the country and would continue for the next decade or so.

One of the last places where steam engines did regular service in GB was in the area around Carlisle in northern England. The locomotives that continued to do battle with the sharp and gruelling gradients to the north and south of Carlisle on the

line between London and Glasgow included some ex-London, Midland and Scottish Railway engines and those of the British Railway standard design of 1951. Most of the ex-LMS locomotives had been built under the direction of Sir William Stanier and included the 4–6–2 'Coronation Pacifics' and some 4–6–0s, nicknamed 'Black Fives' which had been first introduced in 1934. In their heyday, these engines had been regarded by many as the best general service locomotive in the country and in fact their design was the basis of a group of mixed traffic 4–6–0s that British Rail had built in the 1950s.

There were many people at this time, however, who began to take definite steps to ensure that steam locomotive power was kept alive, and not just in memory. To this end a new aspect of railways emerged, the preservation movement, which took two main forms. One was the opening of railway museums where old locomotives and other railway equipment was carefully restored to former glories for perpetual display. The other was perhaps even more positive – the forming of societies who bought old sections of track together with old locomotives. These they restored not merely externally, but to full running order.

The first railway to be thus resurrected was the Talyllyn Railway in Wales, which staged its first run on Whit Monday, 1951, over its narrow-gauge tracks. It is probably less well-known now, however, than the Ffestiniog Railway in North Wales, the narrow-gauge, slate-carrying lines of which were dug out from the wilderness of vegetation that had grown over them, shortly after the Talyllyn Railway had begun its operations. Today the Ffestiniog Railway is nationally famous with regular trains pulled by steam locomotives operating between the coast and mountains of Snowdonia. Even for those with no more than a desultory interest in steam locomotion, a trip on the railway is memorable for its magnificent views of Snowdonia, the historic Harlech Castle, the North Wales coast and the afforested valley of the River Dwryd. The present terminus is at Dduallt, but work is in progress to build new tracks round the lake which flooded the old route, so that the line can be continued on to Blaenau Ffestiniog. Among other locomotives, the railway operates two engines of the Fairlie type,

Above: Locomotive No 323, now known as the 'Bluebell' was built in 1910 for the South East and Chatham Railway and is now in regular service on the Bluebell Railway in Sussex.

Left: One of William Stanier's 'Black Fives', built for the London, Midland & Scottish Railway in the 1930s. More than 850 of this type of engine were built and the design was so successful that British Rail used it as a basis for a class of mixed traffic 4-6-0s produced in the 1950s. The 'Black Fives' were still in operation long after steam had largely been replaced by diesel and electric forms of tractive power.

which were developed from the original 'Little Wonder' – first built and tested on the Ffestiniog Railway in 1869. Wales now has several small scenic railways where old steam engines run.

Moving from Wales into England, the first standard-gauge line to be opened by a preservation Society was the Middleton Railway in Leeds, in 1960. It was followed later the same year by the famous Bluebell Railway in Sussex, which began operations just two years after the line had been closed by British Rail. It has a five-mile stretch of track between Sheffield Park and Horsted Keynes, in the heart of the Sussex Weald. In 1960 the Bluebell Railway began with two locomotives and two coaches; today it can boast a stud of sixteen locomotives dating from 1872 to 1951, although not all of them are in regular use. Besides restoring its locomotives to their original running condition, the Bluebell Railway is anxious to preserve the authenticity of its two stations, for example, by using oil lights to illuminate the platforms.

These then are just some of the places where steam enthusiasts can go to see their beloved locomotives in active service, but the highlight in every steam enthusiast's diary in recent years must have been the 150th anniversary celebrations of the first public railway in the world – the Stockton to Darlington line. These took place in August and September 1975, and besides a magnificent display of old locomotives, some of which were seen chuffing along the railway's original route, there was a mass of related and peripheral activities in the form of exhibitions, speeches and a re-enactment of the laying of the first rail. The first engine to run on the Stockton to Darlington line in 1825 – Stephenson's 'Locomotion' – did not take part, but a fully working replica was built especially for the occasion. The celebrations can only have endorsed the words of Mr R. Janks, who was the chairman of the Western Area Board at the time of the naming ceremony of 'Evening Star', the last steam engine to be built by British Rail. He said, 'Those of us who have lived in the steam age of Railways will carry with us always the most nostalgic memories'.

Right: The 'Earl of Merioneth', pictured here, is one of the two locomotives of the Fairlie type currently in service on the Ffestiniog Railway in Wales. It was developed from the Fairlie engine 'Little Wonder' built for the Ffestiniog in 1869, thus being one of the first articulated type locomotives ever.

Below: A line-up of preserved steam locomotives at the Shildon Works preparing for the 150-year anniversary celebrations of the opening of the Stockton and Darlington Railway in 1875.

Above: A line-up of locomotives ready for action at the Le Mans locomotive terminal.

Left above: A Baldwin-built 2-8-2 'Mikado' at Le Mans Station in France, where until recently they were in regular use. 'Mikados' were introduced into France on the Paris, Lyons & Mediterranean Railway in the first years of World War I.

Left below: A '141P' 'Mikado' pulling an express train and drawing away from Le Mans. These locomotives were designed by André Chapelon and incorporated four cylinders instead of two. Greater efficiency was thus made of the steam as it was used twice instead of once.

France

The 34,000 or so miles of railways in France are closely akin to a spider's web, with the network of lines emerging radially from Paris at the centre. The only major lines that do not begin in Paris are those at the northern foot of the Pyrenees, which comprise some of the mountain rail routes for which France is justly famed.

France's first 'plateways', like those in the early railway days of Britain, were built as long ago as 1782, and a 13-mile line was opened between St Etienne and Andrézieux in 1828. This was to be the first railway on the continent of Europe to use steam locomotives.

Unlike some other European systems, French Railways were at their height of operation before World War II. Since then considerable passenger traffic has been diverted into road transport and much of what remains on the railways is concentrated on to the major lines only. Some lines are kept open solely for goods traffic, while many others have been closed altogether.

Although steam locomotion may have come comparatively late to France, French engineers have been responsible for many important inventions and developments in steam engines. The railways of France have seen the same diversification of locomotives encountered elsewhere in the world, with engines modified and developed in accordance with the demands of rail travel at any given time. Before World War I, the fastest trains running in Europe were in France. The big-wheeled 4–6–2 'Pacifics' found much favour and the French also developed the 4–8–2 'Mountain' type. One of the most outstanding steam locomotives of all time resulted from the inspiration of a French engineer called André Chapelon. Produced in the 1930s, at a time when many French railways were being electrified, it proved that steam locomotion was still very much a viable form of transport. In fact it was a rebuild of a 'Pacific' type locomotive but with many significant new factors, possibly the most important of which was the internal streamlining that facilitated the free passage of vast quantities of steam.

Chapelon's designs were also incorporated into 2–8–2 wheel arrangements and in spite of the fact that many were built elsewhere and imported into France, they were doubtless responsible for the protracted use of steam locomotion in that country. Until quite recently an impressive display of them could be found at Le Mans, a place more usually linked with fast transport of a different nature!

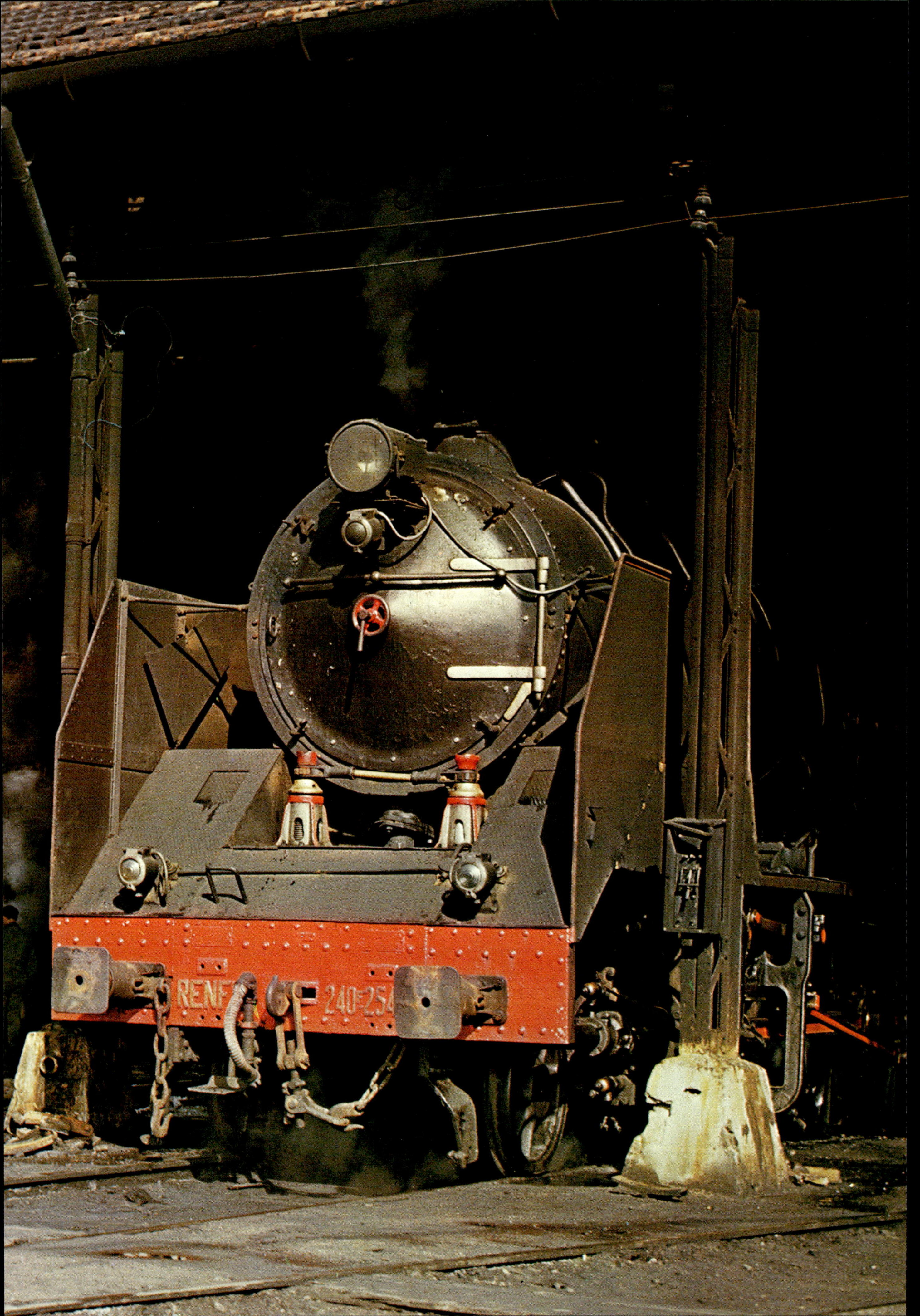
RENF
240-254

Spain

As railways became operational throughout Europe in the nineteenth century, it was clear that they would be of special importance to Spain. Her network of roads at the time was quite inadequate in the linking of place with place and her waterways were unsuitable for regular and efficient traffic. But railways were slow to develop for the terrain, with the steep fall of the Iberian plateau, combined with the country's heavy autumn rainfall, to make for difficult and expensive railway building. Development was further impeded by prolonged governmental indecision, particularly in deciding what gauge to adopt. The standard 4 feet $8\frac{1}{2}$ inches prevailed throughout Europe with only a few exceptions. But the Spanish railway engineers favoured a wide gauge – a choice they claimed to be neither easy to make or irresponsible! Spanish railways therefore, although initially built by small French and English firms, have always had a 5 foot 6 inch gauge which affects the connection with France in lines built across the Pyrenees. Portugal, neighboured only by Spain, was more or less obliged to follow Spanish nonconformity and also built her railways to the broad gauge.

Spain's first railway was opened in 1848 and ran from Barcelona to Mataro. It was soon followed by a rail connection between Madrid and Aranjuez. In 1856 the mountainous 400-mile route from Valladolid to the French border was begun – a line that presented enormous engineering difficulties, dotted as it had to be by bridges and viaducts and tunnels. By the time the Civil War broke out in 1936 there were about 20 individual railway companies which were to merge together to form the Rede Nacional de Los Ferrocarriles Españoles. Railway building is by no means complete in Spain and many of the important connections have been made quite recently. Others are still under construction.

The wide gauge of Spanish railways led to a wide-bodied, massive-looking steam locomotive, happy to traverse the plains but well-equipped to cope with the steep inclines of the mountain passes. The 2–8–0 and 4–8–0 wheel arrangements, which place the locomotive's weight squarely on the drive

Left: One of the massive 4-8-0s widely used over the RENFE *system in Spain. These locomotives were first introduced on the Madrid, Zaragoza and Alicante Railway in an enlarged and improved design in 1920 and were built by Manquista of Barcelona. They are extremely powerful and robust – necessary features for a locomotive which is called upon to haul heavy trains through the stark, rocky cliffs of Spain's mountain passes.*

Below: A 4-8-0 pulling the Zaragoza–Madrid mail train through the canyon of Rio Jalon.

wheels to give greater traction, proved more successful than the 'Pacific' types for heavy general work. 'Pacifics' were widely used throughout the rest of Europe and Spain was almost alone in not adopting them. The 4–8–2 'Mountain' types were used extensively on the heavily graded line from Madrid to the Biscay coast, and at the time of their introduction they were among the heaviest passenger engines in Europe. Many of Spain's stalwart steam locomotives can still be seen in regular use over a wide area of the RENFE system.

USA–Rio Grande

The citizens of Denver in Colorado first chartered a rail line that they hoped would ultimately link them with Mexico, in 1870. As we have seen in an earlier chapter, its original route was blocked by the Atchison, Topeka and Santa Fé Railroad, the first tracks of which were laid in the same year, but the initial transcontinental ambitions of the D & RGW were anyway diverted by the discovery of gold and silver in the Colorado Rockies. With the aim now being to reach, tap and remove the mineral

wealth, a 3 foot gauge track was laid into the mountains, and the 'Rio Grande' became the exception to the recently introduced rule of the United States to build all railroads to standard gauge.

The 'Rio Grande' has been described as the 'scenic line of the world' because of its breathtakingly dramatic scenery of deep canyons and wild mountain passes. The building of the track must be both an engineering nightmare and masterpiece, blasted as it is for many consecutive miles out of

Opposite: Craggy mountain scenery for this Spanish 4-8-0.

Below: Two 2-8-2s built by the Baldwin Locomotive Works in 1925 for the narrow gauge section of the Denver and Rio Grande Western Railroad are seen here outside the engine house at Chama. These engines had a high tractive effort, which was necessary for the heavy gradients and sharp bends of the track. They were originally put to work on freight trains between Salida and Gunnison – the section of line which includes the famous Marshall Pass with its long 1 in 25 gradient.

solid rock. In many places the rugged hills were inaccessible by any other form of mechanical transport, and the curves and gradients, such as the great horseshoe curve at Los Pinos, were tremendous.

Locomotives that found favour in the early days of the Rio Grande were 2–8–0 'Consolidation' types built by the Baldwin company. They proved so successful in fact that they became a standard class, and with various subsequent modifications and additions remained in service for many years. Some thirty years later the Rio Grande could boast some very large articulated freight engines of the Mallet type which proved equal to the demands of the fierce gradients, while passenger services were worked by powerful 'Pacific' type locomotives. In 1938 Baldwin produced some huge 4–6–6–4 'Challenger' types for the Rio Grande in answer to the railroad's demand for a locomotive to pull freight trains through the mountain passes at something approaching passenger speeds, while still being able to cope with the steep gradients.

The years from 'Consolidation' to 'Challenger' saw some notable changes on the actual railroad too. The narrow-gauge system was gradually converted to the standard gauge and as it was then able to connect with other lines at its eastern and western extremities, it became part of one of the great transcontinental routes of the United States. The mineral resources had dwindled considerably by the beginning of the 1900s thus putting different, more conventional demands on the railroad – demands to which the standard gauges was better suited. The first decades of the twentieth century saw a main line of standard gauge stretching the 745 miles from Denver to Salt Lake City, and by the 1920s about 800 miles out of a total of 2,562 was all that was left of the narrow gauge. This figure was further reduced to just over 550 by the end of World War II, and to a mere 260 by 1956. The narrow-gauge tracks that remained were the lines from Alamosa to Durango, Colorado, and from Durango north to Silverton and south to Farmington in New Mexico.

The Silverton Line had a *raison d'etre* in that the country through which it ran was exceptionally beautiful and the Silverton train became a great tourist attraction. The line to Farmington, however, could boast no such reason for existence and according to some authorities it was due for tearing up. Its reprieve came in the form of another discovery of an underground resource – that of natural gas near Farmington, and the removal of this depended on the transport of loads of pipe to the area. Such roads as there were were ill-equipped to cope with the heavy vehicles that would be needed to carry the pipe, so the task fell to the old narrow-gauge railroad. Baldwin built some special 2–8–2s of the 'K-36' Class, which had a high tractive effort. Unassisted they could haul trains of

Above and above right: Locomotives of the 2-8-2 type are seen sweeping round the great horseshoe curve of the Denver and Rio Grande Western Railway's line at Los Pinos in the rugged hills of New Mexico and Colorado.

Right: One of the Baldwin-built 2-8-2s of the Denver and Rio Grande Western Railway thunders along one of the rare, flat, straight stretches of track on this line.

230 tons, but they were often required to work in tandem, for the more usual weight of a single train load was 600 tons. There was one notorious pass with a 1 in 25 gradient, where trains had either to be split into two or three sections, or else three engines – two at the front and one banking the rear – were needed to haul each train. Besides the 'K-36' 2–8–2s, other 2–8–2s which had been converted from standard-gauge models were used.

In spite of the revival boom on its remaining original lines, Rio Grande seemed anxious to do away with the narrow tracks altogether, claiming that to operate them at all was uneconomic. Towards the end of the 1960s, not only had all the gas pipe been laid, but new, more substantial highways had been built to cope with heavy motor traffic. The narrow tracks now carried only one train a week, and that during the summer only, for the snowy winters definitely made it too expensive to keep the line open throughout the year. In 1968 only four freight trains ran the whole year and the locomotives began to find their employment coming from the Hollywood movie-makers rather than the transporting companies. That year brought the closing of the line for active service. Happily, however, the states of New Mexico and Colorado bought 67 miles of the track from Chama in New Mexico (the home of D & RGW's old enginehouse) to Antonito, which they leased to a tourist operator. Under the new name of Cumbres and Toltec Scenic Railway, the line was re-opened as a tourist attraction to the delight of steam-locomotive enthusiasts.

Rio Grande
483

Above: One of the National Railways of Mexico's 4-8-4s retained to run to interchange points with US railroads at the border.

Right: A 4-8-4 built by Alco in 1946 for the National Railways of Mexico, leaves the Valle de Mexico yards on the outskirts of Mexico City. Still operational way into the 1960s, they were the last US style 4-8-4s to work heavy freight trains in mountainous country.

Left: One of the last runs ever of a freight train on the narrow gauge tracks of the Rio Grande.

Mexico

Railways really began in Mexico in 1873, with a 205-mile line that ran from Mexico City to Vera Cruz, although a few localized sections of the track had been in use before this date. During the 1880s, some 6,000 miles of railroad were constructed, stretching from the capital of Texas and forming a link between the two oceans through the Interoceanic Railway of Mexico. This intensive construction was even more remarkable when you consider that the terrain with its high central plateau and steep flanking ridges close to the sea made railway engineering no easy task. Today sections of the Interoceanic Railway rate among the highest railways in the world.

Unfortunately the period of rapid construction and expansion was followed by a period of political unrest in which the railways deteriorated markedly. Conditions did not improve until after the railways were nationalized in 1937, and in fact it is in comparatively recent years – from 1955 to 1965 – that certain Mexican lines have been improved, while other new ones have been built. Nowadays,

nearly all the railroads in Mexico are government owned, by far the largest of them being the National Railways of Mexico.

By 1960, when steam locomotion virtually belonged to a bygone era in the rest of North America, the National Railways of Mexico still had over 400 steam locomotives in regular operation. Many had been purchased from American railroads further north who no longer had any use for them. The 'Star' locomotives were a class of 4–8–4s that serviced the southern part at least of the busy trunk lines that ran from Mexico City up to the border with the United States. They were among the last 4–8–4s to work heavy freight trains through mountainous country.

Above: A strange relationship – a 2-10-0 originally used by the Great Western Railway in Colorado to haul sugar beet and more recently operated on the Strasburg-built railroad in Pennsylvania double-heads with a 4-6-2 owned by the Canadian Pacific on a fan trip in New Jersey.

Opposite above: The same locomotive pictured above on a Toronto–Orillo fan trip takes the bridge at Painswick.

Opposite right: This massive 4-8-4 was built by Montreal Locomotive Works for Canadian National in 1942. It is seen here returning to Montreal after running a fan trip on the Central Vermont.

Canada

Canada has an illustrious railway history that began in 1836 with the opening of a short line called the St Lawrence and Champlain. It was later engulfed by the Canadian National Railway, which is now the largest public-service system in the country. The first trains on this line were pulled by horses, but they were soon replaced by a steam engine imported from Britain and nicknamed the 'Kitten'.

The most famous railway in Canada, however, is probably the Canadian Pacific which was the first line to provide a link between the Atlantic and the Pacific. Built in the latter part of the nineteenth century, its construction in sheer terms of engineering magnitude and human endurance rivals that of America's Union and Central Pacific Railroad. It crosses the precipitious ranges of the Rocky and Selkirk Mountains, and from thence across approximately a thousand miles of prairie and round the rock-bound shores of the great lakes. One contemporary writer said of it, 'Every foot of the mountain divisions of the road was contested, and probably every mile of tunnel and track was sealed with the blood of men.'

Canada now has three main routes from the Atlantic to the Pacific – the Canadian Pacific, the Grand Trunk and the Canadian National. Steam has largely given way to diesel in modern times, but the Canadian railroads have done much to keep steam locomotive power alive by running regular fan and excursion trips using the old stalwart locomotives.

Left, above and right: One of the most spectacular steam operations of recent years has been that of the General Belgrano Railway in Northern Argentina. In eight miles, the track rises from 4000 ft in the lower canyon of Rio Toro to a height of 14,000 ft. These 2-10-2s were the main type of steam locomotive retained to haul heavy freight trains on the northern part of the line to Bolivia. They were built over a period of time by Baldwins and a Czechoslovakian firm called Skoda.

South America

South America has some of the most spectacular railroads anywhere in the world – particularly among the lines that cross the notorious Andes. In addition, South America can claim the highest standard-gauge track in the world – that of the Central Railway of Peru, a branch siding of which rises to a height of 15,844 feet above sea level.

The first railway built in South America, however, was a six-mile line in the Republic of Argentina. It ran from Parque to Floresta in the province of Buenos Aires and was opened in 1857. There is an interesting story attached to the first engine used on the line – it had originally been built for Indian railways and was made for a 5 foot 6 inch gauge. Later it was sent to the Crimea for use during the Crimean War, after which it was bought by the contractors who built the first Argentine line.

Above and opposite: The clouds of black smoke puffed out by the 2-10-2s give testimony to the great effort they have to make as they haul mixed and freight trains along the metre-tracks of the General Belgrano Railway. This is Argentina's most extensive railroad – more than 8000 miles long – and it passes through incredibly rugged terrain as can be seen from these photographs.

This is why many miles of railway track in the Republic of Argentine have been built to a 5 foot 6 inch gauge.

Perhaps one of the most spectacular railways in Argentina is the General Belgrano Railway, situated in the north. In fact this is Argentina's largest railroad system built to a metre gauge. It runs north to the borders of Bolivia and also crosses the Andes to link with Chile. Many people consider the most impressive line of the General Belgrano Railway is the one that runs west through the mountains from Salta to link with Chile at Socompa.

A major feat of engineering, the line traverses tortuous mountain passes and narrow river canyons through a seemingly endless series of loops, tunnels and huge cuttings. After dropping from an altitude of 14,000 feet down to a little town in the Andes that is still situated way above sea level, the line climbs up again to cross the famous bridge, the Viaducto Polvorillo. This line has been serviced for many years by 2–10–2 steam locomotives.

Brazil has a very small mileage of railroad for the size of the country. In 1960, it could only claim 23,750 miles of track throughout its $3\frac{1}{4}$ million square miles, much of it only narrow gauge and single track. By that year also, only some 1,560 miles had been electrified. Railway engineers have always been faced with massive problems mainly arising from the physical nature of the 2,500-foot high escarpment of the Serro do Mar which, although not particularly high, frequently resulted in difficult building operations. It was not until 1945 that a road-railway bridge was opened across the River Uruguay to form rail links with Argentina and Uruguay.

Although its railways are comparatively small in

Below: A Baldwin-built 2-6-6-2 'Mallet' is seen here hauling a train load of empty coal trucks on the Estrada de Ferro Dona Teresa Christina Railroad in Brazil. This railroad has been in operation since the 1880s, when it was driven inland from the Atlantic coast port of Imbituba after coal deposits had been found.

Opposite: A 2-10-4 belonging to the same railroad. A fleet of these locomotives were mainly responsible for hauling coal from Criciuma and Treviso to a coal washing plant at Capivari, and thence to Imbituba. The one pictured here is nearing the port – the Atlantic Ocean can be seen in the background.

mileage, Brazil possesses a line that must have one of the most romantic names in the world – the Estrada de Ferro Dona Teresa Christina, now part of the Rede Ferroviaria Federal SA. Built between 1880 and 1884, it was named after the wife of Don Pedro II, who was then the emperor of Brazil. It runs from the coast port of Imbituba inland to the mines of Lauro Muller, and its main purpose is to transport the mined coal to the coast for shipping to Rio de Janeiro. The steam locomotives that have been responsible for this task are a fleet of 2–10–4s built by Alco and Baldwin, which are supplemented by heavy articulated 2–6–6–2s and smaller 2–8–2s.

Top: Steam locomotives doublehead to haul a heavy freight train on South African Railways in the Orange Free State.

Above: A South African Railways '15F' 4-8-2 mixed traffic locomotive. This type of locomotive, built around the end of the 1930s, was used for passenger or freight work and was a universal favourite throughout South Africa. Great pride was taken in their appearance as can be seen by the immaculate condition of this one at the Bloemfontein workshop. Extremely large engines, the coupled wheels are 5ft in diameter.

Right: A South African Railways 4-8-2 of the '15A' class on the upgrade to Groenbult. Eight-coupled locomotives were found necessary from an early date to cope with the severe gradients on many of the country's main lines, particularly in the Hex River Pass and on the Natal Main line, and the '15A' marked a development from the previously used 4-8-0s. They were fast running engines, quite able to reach and maintain the maximum speed of 55 mph permitted on the steam operated lines of South Africa.

Africa

Railway building in the continent of Africa has a tortuous history and even today the facilities are comparatively poor. To give some indication, statistics produced in 1970 gave the figure of 14,805 miles of railway servicing the 5 million square miles of tropical Africa, whereas Britain had 51,000 route miles in an area of a mere 88,000 square miles. The railway network that exists throughout Africa is furthermore made less effective by the variety of gauges. Much of it was cheaply built and the lines meander extensively, following the topography of the country and using every natural ravine and ridge – rather than tunnels, bridges and cuttings – to gain or lose altitude. Consequently, even in areas that hardly rate the description mountainous, very steep inclines are common. Since the early days many of these pioneer surface lines have had to be realigned and reconstructed to heavier standards.

Railways entered the continent late compared to other parts of the world. The first line of all was from Alexandria to Cairo and was opened in 1854. Five years later, railways made their debut in South Africa, with a two-mile line between Durban and the harbour at The Point. The tracks were laid to the standard 4 foot 8½ inch gauge, but when construction reached the plateau edge, engineers realized that the more flexible 3 foot 6 inch gauge would be cheaper to build and better suited to the terrain that had to be penetrated. The original tracks were ripped up and replaced by those of the narrower gauge, which became the standard measure for South Africa.

Expansion of railways was slow. By 1865, there were just 68 route miles in South Africa and ten years later, this figure had only increased to 155. It was the discovery of mineral resources – diamonds and gold in particular – that was chiefly responsible for their expansion. By the time the Kimberley diamond fields had been opened in 1855, some 1,820 miles of line had been laid, a figure that was nearly doubled in the ensuing ten years, which saw the opening also of the Rand goldfields. In South Africa, Johannesburg became the inland centre to which many railway lines were built from the main ports and by 1892 there were rail connections between this city and the Indian ocean ports of Lourenco Marques, Durban, East London and Port Elizabeth. There was also a circuitous route from Johannesburg to Cape Town. The routes from the Indian Ocean ports had to climb from the low-lying subtropical coastal plains up the 4,000-6,000 foot escarpment that led to the dryer plains of the Highveld. The line from Durban makes its climb near Pietermaritzburg; that from Lourenco Marques near Waterval Boven. Some 85 per cent of the railway's traffic nowadays comprises freight and goods, mainly in the form of agricultural produce

Above: Heavy duties make it necessary for locomotives to doublehead. Here the partners are a '19D' class and a '15CA' class introduced into South Africa in 1937 and 1926 respectively. They make an impressive silhouette against a still sunset sky over the highveldt.

Below: A South African passenger train hauled by a 4-8-2 locomotive chugs up the steep gradient outside Goudplaas. It is en route from Hoedspruit to Pietersburg.

Left above: An example of the '15CA' 4-8-2 mixed traffic locomotive introduced into South Africa in 1926, for passenger and freight duties on the mountainous sections of the main line system. Although brought in to operate on a narrow gauge, they were as tall as locomotives used in Great Britain and generally much wider.

Left below: The 13.15 Mossel Bay train photographed near Letskraal in August 1975. The pilot engine is a class 19D 4–8–2, no. 2643, and one of the last of the 19B class 4–8–2's is the train engine.

and minerals – especially coal, coke and iron ore. Many lines have been electrified and dieselized, but steam still plays a useful part.

In spite of the narrow-gauge track, the early South African engines were usually far longer in wheelbase than those in Britain. This was imperative because of the steep gradients of the track; and operations in Natal, for example, in the early days involved the use of ten-coupled tank engines of great size. In other areas locomotives would be assisted by 'bankers' or 'helpers' which meant an extra engine was either attached to the rear of the train to give extra pushing power, or coupled ahead of the train's locomotive to supply extra hauling power. The most popular types of locomotives were the 4–6–4 'Pacific' and 4–8–2 'Mountain' types. The 'Pacifics' were generally used for the ordinary and express passenger trains, the 'Mountains' were mainly

Above: A lighter 4-8-2 pulls a passenger train on the line from Mossel Bay on the Indian Ocean to Rosmead in the Great Karoo.

Right: These two engines – a '24' class 2-8-4 and a '19R' 4-8-2 doublehead to make the Lootsberg Pass a less formidable task.

Left above: A South African Railway 4-8-4 steams past one of the Three Sisters on the Kimberly-Beaufort West line of the Great Karoo.

Left below: A 4-8-4 traversing the vast barren wasteland of the Great Karoo. This class of locomotive has for years been the dominant motive power on this line.

Above: A 2-10-2 and a 4-8-2 display great effort as they haul an empty train uphill from Umlawula. Happily the wagons are loaded for the downhill run.

Opposite above: A '24' class 4-8-2, shiny and resplendent, heads towards Rosmead on the Mossel-Bay to Rosmead line.

Opposite below: One of the Canadian-built 4-8-2s of the '700' class used on the Swaziland Railway. The line was built in the 1960s to tap iron ore deposits discovered in the mountains of Western Swaziland.

employed to haul goods trains and expresses on some of the hardest sections of track. Garratt articulated locomotives – the largest were 4–8–2 + 2–8–4s – found favour in some places as did some modified articulated Fairlies.

One of the lines to retain steam locomotive power the longest has been the one from the border of Mozambique to the northern Transvaal. When the political problems with Rhodesia and the development of nearby copper reserves forced extra traffic on this line, much of the increased demand was met by its electrification and the dieselization of the line. Steam, however, relieves diesel part way along the track and pushes on to the forest area of the Transvaal Lowveld, before the scenery changes to hills and bush as a forewarning of the climb ahead. The most testing climb is a 16-mile stretch, full of twists and bends, which leads to the Highveld. Various classes of 4–8–2 'Mountain' types are the backbone of the steam operation here and usually two are worked together. Large steam locomotives that used to work the whole route now service this part of the line between Hoedspruit and Pietersburg, where the increased demands for service have been met by laying heavier tracks to support heavier locomotives.

Neighbouring Swaziland is another outpost of the steam empire and conventional steam locomotives may still be found doing regular service over a 3 foot 6 inch gauge track that began its existence as recently as 1964. The railway was built following the discovery of reserves of iron ore in the mountains of Kadake in western Swaziland and the main purpose of the line is to convey this to the Indian Ocean port of Lourenco Marques in Mozambique. It is also used to import goods into Swaziland.

Most of the locomotives on the track are owned by the national railroad of Mozambique – Caminhos de Ferro de Moçambique – which also supplies the train crews. The 4–8–4 'Mountain' types, which are mainly Canadian built, work the upper section of the line, the steep inclines of which rise through a height of 3,500 feet. The lower section of the line down to the port sees a variety of locomotives of which American- and German-built 2–10–2s are the predominant class.

THE REPLACEMENT-DIESEL AND ELECTRIC

The first few years after nationalization of the railways in Great Britain found them heading towards financial deficit. The 1953 Transport Act was introduced in an attempt to counteract this in some measure by giving the Chief Regional Managers, in theory at least, the freedom of setting their own fare charges. This, it was thought, would promote competition and thus efficiency. Plans for major technical progress were still under the control of the central organization and in 1955, one of the most important events in the whole of railway history in Britain was announced – the Modernization Plan. Over the ensuing fifteen years, £1,200 million was scheduled to be spent on 'revolutionizing

Above: The railways of Canada were among the first in the world to change to new forms of tractive power. Here one of Canadian National 'Rapido' trains leaves Toronto Union Station.

Above right: A huge Union Pacific custom-built 8500hp gas turbine locomotive hauls a freight train across Wyoming.

Right: A Santa Fé passenger train – the 'Grand Canyon'.

Far right: The 'flat nose' Alco PA diesel introduced on the Delaware & Hudson line in 1946 runs along the shores of Lake Champlain.

Left above: An early diesel on the New York Central Railroad.

Above: A British diesel hydraulic locomotive pulling an express passenger train in 1968.

Left: The passenger train travelling across the Raritan River in New Jersey is operated in push-pull style with a diesel electric locomotive at either end.

Below: Three diesel units belonging to the New York Central pull empty three-level car-carrying wagons.

Right: One of the first electric engines ever was designed in the 1890s by Thomas Alva Edison, who is seen here standing in the cab.

the railway services'. Steam as a means of motive power would be replaced, and its immediate successors would be diesel and electric tractive power.

Internationally, and even in Britain, neither of these forms of power was new to railways. Indeed a German, Werner von Siemens, had given a demonstration of electric traction as a means of rail transport at the 1879 Berlin Trade Exhibition. By the beginning of the 1890s, electric locomotives were operating regularly in London's underground railway system where their cleanliness in operation gave them undisputed superiority over steam power. Above ground, the Mersey Railway was electrified around the end of the nineteenth century and the North Eastern, having lost a considerable percentage of its passengers to the electrically operated street tramways, followed suit by electrifying its branches north and south of the Tyne. The Lancashire and Yorkshire Railway electrified track around Liverpool, Southport and Ormskirk and the South London line of the London, Brighton and South Coast (LBSCR), experiencing telling competition from the London County Council tramways, also began to experiment with electric traction in 1909. The LBSCR's operation appears to have been the most successful, and with the recovery of all its lost trade, it extended electrification, so that by the 1930s, its successor, the Southern Railway, had electrified all its main lines. But electrification was by no means the success story that it may sound from these examples and apart from certain suburban services around London, few other companies were stirred into action by the glowing financial reports the London, Brighton and South Coast for one, was able to issue. By 1939 only 5 per cent of the total route mileage in Britain had been electrified, and 80 per cent of this could be accounted for by the London Passenger Transport Board and the Southern Railway.

The United States had an electrified main line in 1895 and some twenty years later, forty-two electric engines were operating the same service over 440 miles of track that had previously been undertaken by more than a hundred steam locomotives. The adoption of diesel power has sometimes been claimed as the salvation of American railways although after its initial introduction in 1925, with 300- and 600-hp shunting engines, it remained stagnant for the next decade. The Monon railroad, however, carried out experiments with diesel-electric engines (diesel engines adapted to generate electricity which is then used as the driving force of the locomotive) in the 1930s when many railroad companies were in a state of bankruptcy. It was realized that heavier steam engines would have to be used to haul the heavy trains needed to make operations viable, and yet the nature of the track

Far left below: Among the first diesel electric locomotives to go into service in Britain were these small units shown at the London, Midland and Scottish's Crewe works.

Far left: A metre gauge Bo-Bo-Bo 2400hp 701 class electric locomotive is on the Rhaetian Railway in Switzerland.

Left: Another of Switzerland's electrically operated trains – the 'Blue Arrow' seen on the Lotschberg Railway.

Below: A Pennsylvanian Railroad electric locomotive pulling the 'Executive' train from New York to Washington.

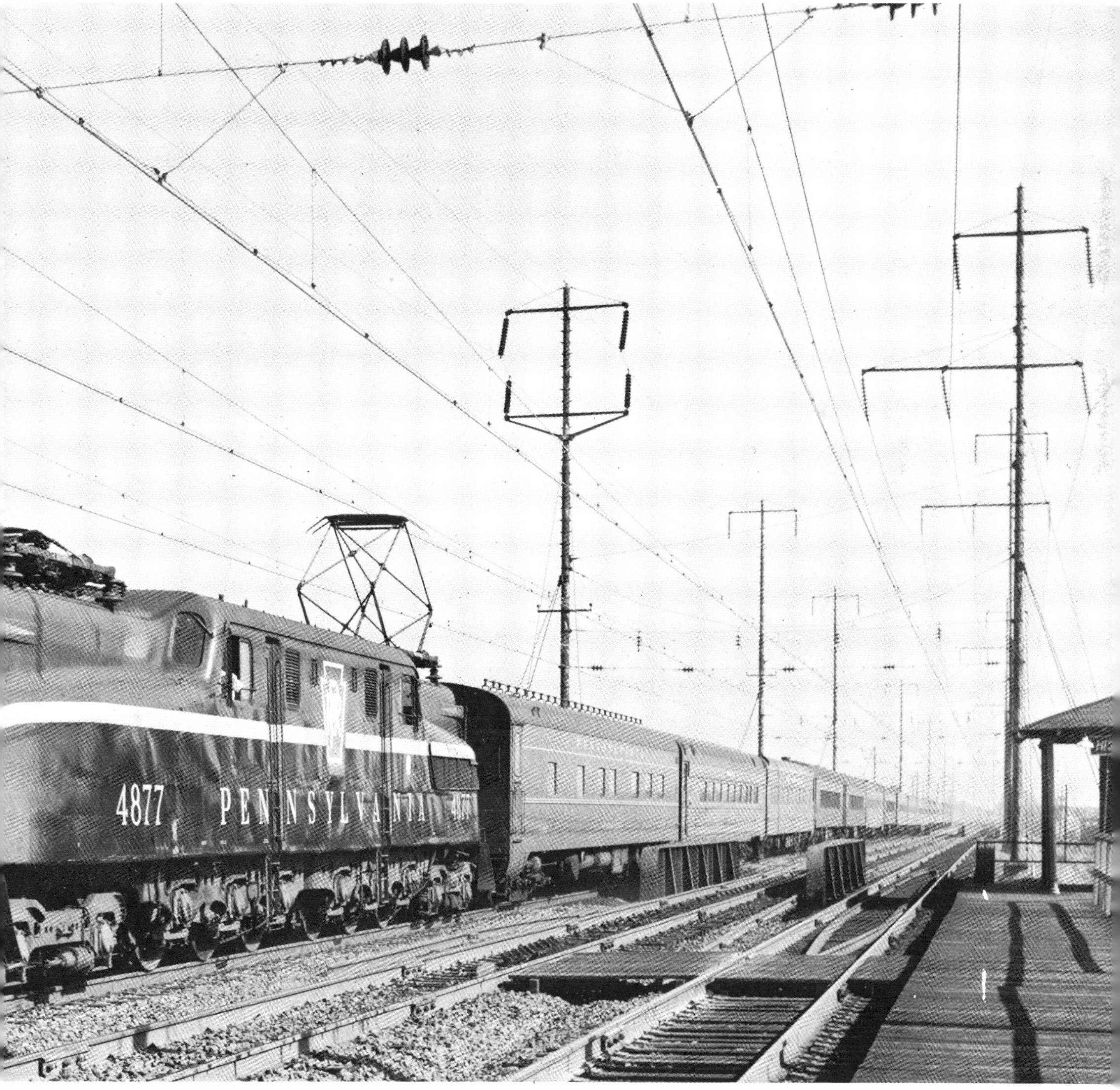

was not sufficient to support them. Diesel-electric engines proved up to the task, and they did no damage to the track. In fact so successful did these experiments prove that the adoption of this form of power spread through the United States railroad companies at an amazing rate over the next few years.

In 1937 bigger line service units of up to 1,800 hp were put into use, and in 1946 United States railroad companies ordered over 800 diesel-electric units as against less than 50 steam locomotives. By 1950 this form of motive power could account for about half the total rail traffic in America, and ten years later, they reigned virtually supreme except where electric traction alone had been retained.

The early days of electrification in Britain found a diversification of ideas about the exact type of power to be used. The London, Brighton and South Coast Railway propelled its trains by 6,000 volts ac obtained from overhead wires. London and South Western who undertook some electrification in the decade of 1910 to 1920 laid a third rail on the track which gave 600 volts dc. Later the Southern Railway adopted the third-rail system and all its overhead sources of supply were converted by 1930. Electric trains are generally operated on a multiple-unit principle, that is one or two coaches with motors are attached to one or two without.

Diesel and electric traction have obvious advantages over steam: rapid acceleration which increases line capacity; less time required for engine upkeep with longer running periods recorded before overhaul is necessary; turntables are rendered obsolete so less room is needed at terminals; less dirt and grime are produced which makes it easier to keep stations clean; and they are generally considered more reliable in operation. Why then with electric traction making its appearance at the turn of the century, and diesel power some thirty years later, were they not adopted as universal forms of rail transport in Britain anyway until well into the second half of this century?

Any innovation that presents a challenge to existing practices and methods will usually be subjected to a period of discredit. Electric traction in particular was seen as a threat to the alliance of coal and steam which held a monopolistic position as the source of all industrial power. Britain, as a major coal-producing country, could probably also see the demands for this product dropping considerably if steam motive power was abandoned, so the railways' 'anti-attitude' was no doubt given support from political as well as business circles. Railway companies, it seemed, went almost out of their way to prove that steam could offer at least as efficient a service. The Great Eastern Railway answered its public's demands to electrify their suburban service in 1902 by building a huge,

Top: The main line diesels operating over France's SNCF are typified by the A1A-A1A- 2700hp 6800 class of locomotives shown here.

Above: General Motors built this 1500hp Co-Co diesel electric locomotive for the 5ft 3in gauge of Victorian Railways.

Right: Mistrust seems to be registered in the very stance of a steam engine driver as he watches a diesel light passenger unit about to go out on a test run from Euston Station in 1938.

ten-coupled steam engine that could achieve a speed of 30 mph in less than 30 seconds, while hauling a train weighing more than 300 tons. The company patted itself quietly on the back, while its passengers waited another fifty years for electrification to arrive.

Apart from the threat to steam and coal, and all this entailed, there was a very considerable cost factor involved! In 1931 the Weir Committee estimated the capital cost of electrifying British Railways at £261 million, and that figure did not include the cost of constructing and equipping generating stations. People could even be forgiven for considering this sum in a somewhat cynical light,

for estimated costs of electrification conversions seem often to be substantially exceeded by the subsequent amounts actually spent. Furthermore, the Weir Committee Report came at a bad time – in the middle of the Depression era, when it may have been harder to visualize a return on such a vast investment.

Diesel-electric traction was a somewhat cheaper alternative to electrification and seemed to possess many of the same advantages. A comparative cost study was published soon after the Weir Committee Report, which estimated that the net costs of conversion to diesel over the same route mileage as the electrification estimate would be about £154 million. The report pointed out that it would also be considerably more economical to operate and maintain than its predecessor, steam. In addition, diesel had the advantage over electric traction of that it could be introduced in separate units, so diesel engines could take over from steam locomotives as they came to the end of their life. A transference to diesel did not involve major track installations as it did with electrification. With these points apparent, and the fact that experience in the United States had proved diesel-electric operations to be not only viable but lucrative, it is hard to say why there was not much reticence in adopting the system. It may have been because of the poor publicity that had been given to both the financial and technical advantages and it may have been a natural, inborn reluctance of the railwaymen – brought up to think only in terms of steam – to admit that anything could take its place.

In the early days of diesel (the first years of the 1930s) it too represented a threat to the coal industry. But none of the opposing arguments could hold up progress for long, and in the late 1950s dieselization was proceeding on British Railways on an impressive scale. By then British manufacturers of steam locomotives were noticing a marked drop in the overseas markets, because the world in general was changing to the more modern forms of motive power. Good steam coal was beginning to be in short supply and thus the price rose sharply. And the attractive, clean and much-improved working conditions that many factories were beginning to offer to employees made the job of driver or fireman on a dirty steam locomotive less appealing than it had previously been.

The main emphasis of the Modernization Plan of 1955 was placed on diesel traction and the immediately ensuing years brought a boom in the manufacture of small diesel railcars for suburban and secondary main-line and branch services. By the end of 1958, when the most important, and most expensive, stage of British Rail's dieselization was launched, over 2,000 of these vehicles were in operation. Only just over 100 main-line diesel locomotives were in use, however, although a period of five years brought this figure up to the 2,000 mark too. By 1965, British Rail owned more than 4,000 diesel multiple units and nearly 5,000 diesel locomotives, which comprised nearly fifty different classes with choices of mechanical, hydraulic and electric transmissions. Such feverish expansion brought considerable problems, which quickly gave the new locomotives a reputation for unreliability. With numerous manufacturers of diesel engines, in addition to the established steam locomotive builders, pressing for locomotive orders to keep them busy as well as to attract export trade, few prototypes had time to be sufficiently tried and tested. Rail travellers grew impatient of the continual apologies of 'it's merely teething troubles'. Their impatience was heightened by the fact that it appeared as if one kind of power – steam – had merely been changed for another – diesel – that seemed to have no advantages while offering a less

reliable service. This false view arose because at first diesels literally just took over from old steam locomotives that had passed out of service, so they had to run to the old steam-locomotive schedules, in conjunction with steam locomotives still in use. Diesel operation would only be able to show its true nature and maximum efficiency when a complete service was dieselized. In addition, many departments of the rail network were simply not ready to operate and maintain diesel locomotives. Whereas steam engines could be kept going by relatively unskilled workers with little more than a basic

Officials and spectators gather to watch the auspicious moment as the first diesel electric locomotive to run on any main line in Britain prepares to leave St Pancras Station bound for Derby and Manchester.

The Union Pacific's reputation for operating huge locomotives, as illustrated in the steam era with the 'Big Boys' and 'Challengers', has been carried over to the replacement forms of tractive power. Here a massive UP – 52 5000hp locomotive used for hauling heavy freight trains is seen at Green River, Wyoming.

knowledge of the mechanics, diesel locomotives need trained technicians to carry out servicing and maintenance programmes, in properly equipped centres. Only Eastern Region seemed to have a full understanding of these needs and it allocated considerable sums of money for building new maintenance and inspection depots. Such foresight was to pay off in 1964 when the most numerous type of locomotive – the Brush Type 2, in use since 1957 – began to develop metal-fatigue problems. By this time, Eastern Region was fully dieselized and mainly

using these particular locomotives. The region's new timetables were based on the assumption that 82 per cent of the fleet would be available for use each weekday and it had no steam locomotives left to take over in times of crisis. Eastern Region's well-trained maintenance staff proved their worth and, with some help from British Rail's central mechanical engineering personnel, it was able to keep all the affected locomotives in service until such time as each one could be re-engined.

It seems particularly ironic that the Brush Type 2 model should have developed such a problem, for it incorporated features used in a series of locomotives built just after the war for the Ceylon Government Railways and thus did have considerable proving and testing behind it. Built with engines of 1,250 hp, 1,365 hp, 1,600 hp and 2,000 hp, it was looked upon for the first five years of operation as one of the most successful designs in service.

The first diesel locomotives in use tended to be unnecessarily cumbersome because of stringent restrictions imposed by civil engineers on axle loading in particular. Non-powered axles were included in the design to help spread overall weight and avoid placing excessive weight on the powered axles. As a result, the first passenger diesel locomotives were underpowered for the purpose of producing higher speeds than steam engines; much of their power in any event was used in merely moving their own bulk. Once dieselization was more fully implemented over the railways, however, the development of diesel-electric traction became rapid. The civil engineers' restrictions became less strict and the non-powered axles were eliminated from many designs. The electrical engineers and engine manufacturers, through increased experience, were able to produce improved generators and motors of higher tractive effort, and achieve an increase of available power per ton and cubic foot of engine.

The diesel locomotive that has possibly best demonstrated the capabilities of diesel traction has been the English Electric 3,200-hp 'Deltic', which derived from a prototype built privately by this company in 1955. At this stage no other manufacturer had been able to produce a locomotive that combined even as much as 3,000 hp with reasonable size and axle loading.

The electrification that came about as part of the Modernization Plan was carefully projected, doubtless because of the heavy costs involved. As we have seen already, these are primarily incurred in the actual installation: the cost of procuring, maintaining, and to some extent running, electric rolling stock is less than for any other method of traction. Unlike diesel locomotives, which cannot produce more than their rated hp power, the power available to an electric train is virtually limitless, restricted only by the current load its electrical equipment can withstand without the overload relays tripping. Electric locomotives rated for normal purposes at about 3,000 hp, generally have a one-hour rating of up to 4,250 hp, and at peak output this can be extended to more than 5,000 hp. They are also claimed to need less routine maintenance than diesel locomotives. And in any case the maintenance needed is less complex to perform.

The first two electrification programmes to be

carried out after the war were between Manchester and Sheffield via Penistone and between London's Liverpool Street and Southend. They were equipped with the 1,500-volt dc system then officially recommended. Shortly after the publication of the Modernization Plan, however, it was decided to adopt as standard the electrification method that employed a 25,000-volt ac supply, which French National Railways had used so successfully between Valenciennes and Thionville a year or so earlier. Southern Region, which was almost entirely electrified, retained its third-rail system.

The main advantage of the 25,000-volt ac system over the 1,500-volt dc system is that it needs much less ground equipment to feed the current to the overhead wires. It also economizes on the weight of these wires, which are smaller in section and therefore lighter and thus need less bulky structures to support them. The ac system also means the railway can draw its current supplies direct from the national grid. It can, however, involve high initial expenditure in the necessary modifications to the overline structure in order to give the contact wire (the catenary) the clearance in requires. This was amply demonstrated on the electrification of the London Midland Region's system from Euston to Liverpool and Manchester, a programme that did much to bring electrification into disrepute in some quarters. Forecast to cost about £118 million, the figure five years later, when it was still not completed and much of the equipment had stood idle for some years, was nearer £160 million.

When this scheme had been completed in 1966 a policy was initiated whereby all locomotives should

Above: A 'box-shaped' Class E-44 electric locomotive used by the Pennsylvania Railroad for hauling freight trains on the East Coast runs.

Far left: Three diesel-electric units pull an eastbound freight train on the Erie-Lackawanna line. The Delaware, Lackawanna and Western amalgamated with the Erie Railroad in 1960 to form this line, which operates independently although it has been owned by the Norfolk & Western Railway since 1968.

Left: An electrically operated block train of specially designed vans runs between Ford Factories at Dagenham and Halewood.

be 'common users' – that is they could be used for passenger or goods work. A number of different manufacturers supplied locomotives, which nevertheless all looked very similar to one another. Eight years later, electrification had been extended to Glasgow and experience with the locomotives had indicated areas where they could be improved. In consequence the more powerful Class '87' was produced with improved tractive ability and smoother riding characteristics. These 5,000-hp locomotives are now used to haul the 'Electric Scots', the fastest of which averages 80 mph for the London to Glasgow run.

The 25,000-volt ac electrification has been extended to the old Great Eastern main line from Liverpool Street to Colchester, to the Glasgow and the Liverpool Street suburban services, and over the London, Tilbury and Southend lines out of Fenchurch Street to Shoeburyness, as well as over suburban services out of King's Cross. Southern Region has electrified services on all its main lines in the South East and from Waterloo to Bournemouth. Authorization for this last-named scheme in late 1964 seems in retrospect to have coincided with the acknowledgment of the benefits of railway electrification; and the increase in passengers on electrified services over the next few years certainly demonstrated its popularity with the public.

Outside Britain electric and diesel forms of motive power have had much the same uphill struggle for acceptance. We have already seen how diesel traction stagnated over a period of ten years or so in America before diesel-electric operation gained widespread popularity. Today diesel-electric traction, using mainly moderately powered units in the 1,000- to 2,000-hp range, handles about 90 per cent of America's traffic. Although the country was a pioneer in electric traction, with the main lines of the Baltimore and Ohio Railroad operating as early as 1895, it has by no means become widespread. In some areas it was more or less forced on the railroads – to counter bad pollution and accidents resulting from the smoke generated by steam, particularly in extensively tunnelled areas. Only two main railroad companies have switched to electrification over any notable distance: the Chicago, Milwaukee, St Paul and Pacific, and the Pennsylvania, which both have over 600 miles of electrified track. Other companies, such as the Norfolk and Western, electrified track over the Appalachian mountains, only to revert to steam sometime later when parts of the route had to be realigned. At the peak of electrification in the United States only just over 3,000 miles – representing little more than one per cent of the total railroad mileage – was electrified, and since that time this figure has fallen considerably. Probably the main reason for this low figure is that the density of traffic on most railroads, particularly over long distances, is simply not sufficient to justify the huge financial outlay of converting the track.

In America and Canada diesel-electric traction is the undisputed king, first gaining a footing in the inter-war years. Both the Canadian National and the Canadian Pacific are almost completely dieselized. Along the north shore of Lake Superior, the Schreiber Division of the Canadian Pacific has increased the weight of trains hauled by using two diesel-electric units per train. Although this does not add to the train crew, as happens when steam is double-headed, it has the effect of either almost doubling line capacity or halving the number of trains required.

On the continent of Europe, diesel and diesel-electric traction made an inter-war debut in Germany and Denmark, but since that time it has become more widespread and in most places now handles the major percentage of rail traffic. Electrification has always found initial favour in mountainous countries, where cheap hydro-electric power abounds and where it is also the most suitable form of traction for lines that are beset with long tunnels and heavy gradients. Switzerland provides the supreme example and began her conversion to electric from steam as early as 1919 using a single-phase 15,000-volt ac method. Locomotives have always been used in preference to multiple units, mainly so that they can be readily interchanged to haul through passenger and freight traffic from other countries. Nowadays over 95 per cent of Swiss railway's route miles are electrically operated, mostly employing multi-voltage electric

Above: One type of locomotive assists with the installation of another. At Liverpool Street Station in 1960, diesel electric locomotives pull the wiring trains for erecting the overhead wiring that will allow electrification of the lines from Liverpool Street to Chingford, Enfield, Hertford and Bishop's Stortford.

traction for international express work. This means that engines can haul Trans-Europ-Express (TEE) trains, for example, over electrified networks of four differing voltages, from 1,500 and 3,000 volts dc to 15,000 and 25,000 volts ac, thus enabling them to provide international services throughout Belgium, the Netherlands, France, Western Germany, Austria and Italy.

Other European countries with a proportionately high percentage of electrified route miles are Austria, Norway and Sweden. Again their mountainous terrain and thus the viability of cheap power had led to them being able to electrify quite early without the careful, and often prohibitive, financial estimates of returns from traffic density alone. As a result they all operate over 60 per cent of their route miles by electric traction. The Netherlands, where two-thirds of the rail network was devastated during World War II, has found it economic in the reconstruction to electrify only about 50 per cent of its route mileage. The other half, which carries only about 25 per cent of the traffic, is worked by diesel-electric traction.

France was active in developing electrification for

Above: This triple-articulated road driven Swedish unit – an SJ class DM3 1-D+D+D-1 – was specially built in 1960 for working 4000 ton ore trains between Kiruna and Narvik in the Arctic Circle.

Below: Since dieselization and electrification became common forms of tractive power on railways throughout the world, France has produced some revolutionary locomotives. Pictured here is the SNCF Quadricourant, which is able to operate over four different voltage systems. It was specifically designed to pull heavy trains on fast schedules.

Right above: Another Swedish train. This one is a Ra class 3600hp Bo-Bo electric locomotive introduced by Swedish State Railways in 1955 for hauling heavy express trains.

Right below: Although conditions vary greatly throughout the Austrian Federal Railways System, they are now almost entirely electrically operated and locomotives are expected to surmount the heavy gradients of some lines as easily as they must maintain high speeds over the main lines. The 3400hp locomotive illustrated is one of the more powerful types, able to pull the 'Transalpin' express the 580 miles between Vienna and Basle in under 12 hours.

RAPID
7
ASEA

Above: Two Norwegian locomotives – a multiple unit for local work, and a 2200hp class Bo-Bo which operates between Oslo and Berger.

Below: A 'double-deck' car used on some of Canadian Pacific Rail's suburban trains.

Opposite: A diesel-electric locomotive on the Denver & Rio Grande Western pulls a freight train through the Tennessee Pass.

Overleaf: Two Canadian Pacific Rail diesel-electric units pull a loaded freight train through the Red Sucker Tunnel, near Port Coldwell, Ontario.

main lines soon after World War II having come into possession of the pioneer electrified line in the Black Forest, which was then in the French zone of occupied Germany. So impressed were the French with the merits of the 25,000-volt ac system of this line that they applied it to their pilot schemes in the French Alps. Electrification on French lines before this had been at 1,500 volts dc, but after various tests in which they eradicated defects and perfected design techniques, they used the higher-voltage method for further extension of electrification.

Two French Railways electric locomotives hold the current world record for speed. The six-axled CC 7107 and the four-axled BB 9004 locomotives both reached a speed of just over 205 mph on successive days in March 1955. These record runs took place in Southern France.

Outside Europe, in countries such as the USSR, South Africa and Australia, steam has mainly been replaced by diesel or diesel-electric traction. The story is the same – the massive costs of electrifying the railways could not really be justified by the density of traffic, and diesel-electric haulage was a more economic alternative. In spite of the seemingly low percentage of electrification in terms of actual route miles, however, it also has to be balanced with the amount of traffic electrified lines actually carry. The USSR, for example, has between 20 and 30 per cent of its network electrified but this actually handles more than 50 per cent of the country's total rail traffic. South Africa is constantly extending its electrification to cope with increasing demands, whereas most of Australia's electrification has been concentrated on suburban lines around Sydney and Melbourne in particular and amounts to only about 2 per cent of the total network. The rest of the country is really still too underdeveloped to warrant electrification. A variety of diesel-electric units are

3079
3079
Rio Grande
3079
3039
Rio Grande

4242
4242
CP Rail
4242

CP Rail
4239
CP Rail

Above: Diesel No 1454 pulls six carriages through the Cumberland Gap, heading for Washington DC on what was one of USA's first public railroads – the Baltimore and Ohio.

Left: An electric locomotive drawing its power from a third rail on the track leaves the Grand Central Station, New York.

Below: Two Santa Fé 'cowl-type' diesels haul a passenger train over the steep grades of Raton Pass in New Mexico.

Right: Spectacular scenery for those travelling on Canadian Pacific Rail's 'Canadian' at Stoney Creek Bridge, in British Columbia.

used over the railways, mainly in the 1,000- to 2,000-hp range, although Western Australia uses more powerful units to convey its heavy mineral traffic. Branch lines in Queensland, South Australia and Victoria with their light traffic employ several types of diesel railcars and trailers.

Japan's performance in electrification is one of the most impressive in the world; its Tokaida line between Tokyo and Osaka provides the prime example. This line is serviced by Japanese National Railway's 'Hikari' or electric 'bullet' trains that provide a 100-mph-average service at 15-minute intervals during the day. Extensions of this line are planned and under construction in many parts. On the newer sections it is estimated that the 'Hikari' trains will reach speeds of 160 mph (as against current top speeds of 130 mph).

There is considerable conjecture, with some clear indications, of just what form future rail travel will take. With emphasis placed on improved speed and performance, presumably combined with economy of operation and passenger comfort, British Rail have had the HST – High Speed Diesel Train – in operation for some while for long-distance travel. It is capable of maintaining constant speeds of 125 mph, but as this is not considered fast enough, tests are currently being conducted on the Advanced Passenger Train. This will be able to achieve speeds of 150 mph and through the tilting action of its body, it is able to negotiate curves in existing track at far higher speeds than are possible with conventional rolling stock.

Some countries have experimented with gas-turbine expresses, which could find further favour if the cost of fuel oil becomes prohibitively high. Gas-turbine expresses run between New York and Boston, and Montreal and Toronto. The French have built a train powered by four gas turbines driving turbo-alternators that provide the electric current for the transmission. Known as the TGV001, it has achieved speeds of about 200 mph on test runs.

Other areas of experiment include overhead monorail systems, hovertrains and linear induction, in which the train is propelled by magnetic forces. The Japanese would appear to be closer than anyone to putting any of these experiments into practice: they have built a magnetic levitation research vehicle. Propelled by linear motors and operating by direct electro-magnetic force, it is capable of maximum speeds of more than 300 mph and has been developed because it is predicted that the 'Hikari' trains will not be able to meet the travelling needs of Japan in the foreseeable future. Instead there are plans to have a one-hour service using the magnetic levitation vehicle in operation between Tokyo and Osaka by 1980.

Right: Tests began on British Rail's Advanced Passenger Train in September 1973. It is hoped to achieve speeds of 150mph without having to make expensive modifications to existing tracks.

Below: SNCF's TGV 001 is powered by four gas turbines driving turbo-alternators which provide the electric current. It is intended to provide a two-hour service between Paris and Lyons, for which it will have to maintain an average speed of 158mph.

Opposite above: British Rail's 'High Speed Train' has been designed to run the major inter-city services on non-electrified lines. Its maximum operating speed is to be 125mph although on test runs it has broken the world speed record for diesel traction by running at 141mph.

Right below: The Japanese 'Hikari' or lightning trains maintain speeds of 130mph or more on their run between Tokyo and Osaka.

Servishell
キリン 
ビール
岡本旅館

RAILWAYS GO UNDERGROUND

As the size and population of the great capital cities developed, it became apparent that there was an increasing need for a new type of railway. This was not so much to cover long distances between built-up areas, as to convey people from one part of the city to another. Nowhere was the need for this better demonstrated than in Victorian London, where the narrow streets seemed to be perpetually blocked by horse-drawn buses and carriages. To build a street railway would obviously only make the congestion worse, and the problem of where to put a city railway was theoretically answered in two ways – it could either be built underground or overhead. The overhead idea was favoured by many people in the 1830s and they visualized a circular railway enclosed in glass running around London. To the advocates of this 'Crystal Way' the idea of an underground, or 'sewer' railway as it was derisively called, was anathema. Protests about airless holes suffocating people or sending them mad were heard, but a more practical and sensible objection was the difficulty of digging and constructing long tunnels safely.

Marc Brunel, father of Isambard Brunel of Great Western Railway fame, was the first person in Britain to construct a tunnel beneath the Thames and the problems he encountered were enormous. Work began on this in 1825, and it was not completed until 1843 following at least two disasters when the murky waters burst through the bricked walls of the tunnel. Not intended originally for rail traffic at all, it was acquired by the East London Railway in 1865 and was used by trains in 1869 on a service from New Cross to Wapping.

London's first underground railway, however, dates back to 1845 when there were 19 Bills deposited proposing railways in the Metropolitan area. A Royal Commission was appointed to review the situation and one of the schemes they considered was that put forward by Charles Pearson, solicitor to the Corporation of the City of London. His

Below: A print depicting an early attempt in Britain to construct a 'tube' underground railway. The 'tube' refers to the method of construction – the earlier metropolitan line had been built on a 'cut and cover' principle, which means a trench was dug and the railway was then built and roofed over. A 'tube' railway was constructed entirely underground by sinking shafts and tunnelling beneath the surface.

Bottom: Rush hour, it seems, never changes! In this Doré print of 1875, workers are crowding to get on the train. The engine was typical of the tank engine types used on the underground system for some years. They had no cab; the driver was protected only by a weather board.

suggestion was for a wide road from King's Cross to Farringdon Street, with six standard-gauge and two broad-gauge tracks in a tunnel below. The Commission reported adversely on the scheme but Pearson did not give up. A later Bill, along similar lines but with considerable changes from Pearson's original, was successful and work began in the 1860s on the first of the covered lines.

It was to be built on the 'cut-and-cover' principle – that is, a trench was dug for the railway itself, which would be roofed over so that the line was completely covered. The roof could later form a road. All the problems that may be imagined seemed to have been encountered. Beset from the start by irate house owners who demanded large sums in compensation for the cracks in their walls which they claimed were caused by the workers below, the engineers decided to dig as much as possible directly below the lines of the existing streets. This brought its own difficulties, for just beneath the surface lay networks of water, gas and sewerage pipes, and also, surprisingly enough at such a time, electric telegraph wires. All these had to be diverted. The line was finally laid with three tracks so that trains of both standard and the GWR's broad gauge could run over them.

The $3\frac{3}{4}$-mile line took $2\frac{1}{2}$ years to build and was opened on 10 January 1863. Known as London's Metropolitan Railway it was the world's first passenger-carrying underground railway and the 'father' of 'Metros' the world over. As the Great Western Railway owned a shareholding in the company, it provided the first rolling stock comprising broad-gauge locomotives and carriages. Relations between the GWR and the Metropolitan company were far from good, however, and by August they were so bad that the GWR withdrew all its rolling stock. The Great Northern Railway and the London and North Western Railway subsequently provided locomotives and carriages until at least the following July, by which time Metropolitan had begun to acquire its own equipment.

Running steam locomotives in enclosed tunnels brings obvious problems and the first locomotives were designed in such a way that steam could be diverted into water tanks instead of up the chimney, with the surplus escaping through small chimneys on either side of the tanks. The Metropolitan's own original locomotives were 4–4–0 type tank engines of the 'A' class built by Beyer, Peacock & Co. In their original form they had no cab – the only protection the driver had was from a weather board, or 'spectacle plate' as it was called. These 'A' class locomotives represented the basic design of virtually all locomotives used on the underground system until electrification.

Following the opening and successful operation

Above: The underground railway that passed close to St Paul's and the 'Times' office is shown in this print of 1875.

Right: Electrification made its debut on the underground at the end of the 1880s, but in 1902, when this photograph was taken at Aldgate Station, the 'Inner Circle' was still using the old fashioned steam powered locomotive.

(an estimated 9,500,000 people travelled on the Metropolitan in its first year) an eastward extension to Moorgate was brought into service in December 1865. Then in 1870 another London underground line opened – the Metropolitan District Railway. It was intended to operate as a partner to the Metropolitan but in fact they were bitter rivals, until 1884 when they combined to form the Inner Circle Route.

London's first 'tube' railway was a subway between Tower Hill and Bermondsey serviced by a cable-operated car and opened in 1870. It differed in construction from the underground railways in that vertical shafts were sunk and tunnels were then driven direct from the shafts entirely underground. The shaft and tunnel were lined with a cast-iron shell or 'tube' to prevent the earth falling in. The narrowness of this tunnel did not make for successful operations, and after a very short time the car was withdrawn and the subway converted to pedestrian use only. In 1896 it was made more or less redundant by the opening of Tower Bridge. Because of this failure it is the City of London and Southern Subway that is more generally regarded as the inaugurator of the modern tube system. It began with a 3-mile line to Stockwell opened in 1890 and the two tunnels it incorporated were both unsuitable for steam locomotives. Originally it had been intended to operate the subway with cable-operated

INNER CIRCLE

TAYLOR'S
DEPOSITORY
PIMLICO
VICTORIA
REMOVALS &
WAREHOUSING
FURNITURE, LUGGAGE
EXPERT SERVICE
15
15

Above: An even older electric engine in use on the underground is this one at Borough Station on the City and Southern London Railway. Although the photograph dates from 1922, the engine dates from the opening of the line in 1890.

Left: The Waterloo & City Line were early users of electric traction. One of the smart coaches designed by Dick Kerr is shown in operation here in 1899.

cars but this scheme was dropped in favour of electric traction which had made its debut a few years earlier. The electric locomotives built for the line were small and pulled two or three cars. These had only narrow slits of windows near the roof, for it was considered that as there was nothing to see while the train was in transit passengers would not want to look out of windows anyway! The carriages were padded from floor to ceiling in leather and were known as 'padded cells'.

The City of London and Southern Subway line was opened by King Edward VII – then Prince of Wales – who turned on the current with a golden key at a special ceremony. He then travelled the length of the track in one of the 'padded cells', while a conductor called the name of the stations.

The construction of the 'tube' was made possible by Marc Brunel's invention of a tunnelling shield which was improved in 1869 by Peter Barlow. Further refinements were later added by Barlow's friend and pupil, James Gatehead. The 'shield' consisted of a series of great iron or steel rings bolted together, which fitted into the first section of the tunnel. As digging progressed, the steel rings were forced forward by pressure and other rings were added. The 'shield' was found to be particularly effective for tunnelling through the London clay.

From that time on, nearly all of London's underground railways were constructed on the 'tube' principle. Early in the 1900s there were seven tube railways – the City and South London which had extended considerably to the north and south, the Waterloo and City, the Central London, the Great Northern and City (Metropolitan) Railway and the Piccadilly, Bakerloo and Hampstead sections of the London Electric Railway. The Central was known as the 'Twopenny Tube' soon after its inception, and it was an extremely busy and popular line, serving as it did the main theatre and shopping areas. The underground system interlocked at many places and different levels and as they extended into suburban areas further out they would emerge into daylight. In fact nowadays only about one-third of London's Underground system is actually underground. Many of the lines were constructed to 'dip' between the stations, so as to give a saving in electric current. The principle of this is as a train leaves a station, which is constructed on a summit, it straight away encounters a down grade which allows it to pick up speed. A run along straight track before rising to the level of the next station acts as a natural brake thus assisting in bringing the train to a standstill. London's Underground can still boast the longest continuous vehicular tunnel in the world – approximately 17½ miles from Morden to East Finchley.

The immediate success of London's urban railways prompted other cities to follow suit. A cable subway was built in Glasgow passing twice under the River Clyde to make a complete circle of the city. Originally the trains, which resembled tramcars, were worked by cables operated by stationary steam engines. Later the line was electrified and in fact became one of the first to have electrically lit carriages worked by current drawn from a central third rail. Merseyside followed in 1886 with an underground railway. It had steep inclines that made the engines work so hard that huge steam fans were installed to clear the air. Liverpool was the only place in Britain to pursue the idea of an overhead railway. In 1893, nearly seven miles of track forming the Liverpool Overhead Railway was operated along the shores of the Mersey. It was the first overhead electric railway in the world. Nicknamed the 'Docker's Umbrella' as it kept dockers dry on their journeys to and from work in wet weather, it was an ugly monolithic structure supported on wrought-iron girders embedded in concrete. Upholders of aesthetic values were gratified some years later when the line was closed so it would no longer present such an eyesore on the skyline.

Above: The 'boxes' on the post to the right of this picture are electronic devices, designed to help control the speed of the train as it approaches Times Square on the New York City Automated subway.

Right: Metropolitan and Bakerloo line trains at Wembley Park. The train in the foreground is one of the Metropolitan 'A' stock 'silver' trains of 1938.

Elsewhere in the world underground railways were also constructed. New York had its first rather clandestine subway in 1867. It was built secretly by Alfred Ely Beach and its 100-yard track ran from the basement of a building on the corner of Broadway and Murray Street. Beach ignored the regulations which had said he could build two 4 foot 6 inch diameter tunnels for pneumatically driven vehicles and instead built a 9-foot diameter tunnel

NOTICE
LOCAL
1923B0
1923C0

UXBRIDGE
3145

NEWARK

601

WEMBLEY PARK
A
STANMORE
BAKERLOO LINE
51

and a pneumatically driven passenger car conveyed people along the experimental line. New York, however, was to meet the need for an in-city railway, in the first instance by constructing an overhead railway, the first section of which was opened in 1878. Subsequently it was the most complete system of this kind in the world. Charles Thompson Harvey's Elevated Railway, or 'EL', extended as far as Bronx in 1886 and thus considerably widened the area of New York. At the peak of its operations it was claimed to be carrying one million passengers each day. At first the railway was worked by steam

Far left top: A train entering Harrison Street Station en route to Newark. This is part of the New York Authority's Trans-Hudson system known as PATH.

Far left centre: A new answer to in-city travel has been put into practice in Chicago. Here cars and trains run side-by-side – the train running on rails constructed on the central reservation of a motorway.

Left top: The Market-Frankford line in Philadelphia is also part underground and part elevated. It uses trains constructed of stainless steel, which are said to be forty per cent lighter than those formerly used.

Left centre: This elegant train with its streamlined look is one of those belonging to San Francisco's BART system.

Left: Pre-war rolling stock at Wembley Park Station belonging to the Metropolitan and Bakerloo lines.

Above: The San Francisco Bay Area Rapid Transit District system – known as BART – operates at various levels. At the lowest level are the BART trains, at the intermediary level – the municipal platforms and the concourse is just below street level. The system also operates at ground level and on an elevated plane.

Right: Automation on London's newest underground system – the Victoria Line.

locomotives, but these were later replaced by electric traction.

Overhead railways in effect fulfil the same function as those underground, although undoubtedly at the expense of the appearance of the city. They are, however, considerably cheaper to build, for columns, girders and arches, unlovely though they may be, cost less to build than complex underground tunnels. The average price of New York's elevated railway when it was built was about £70,000 a mile, whereas the first section of London's Metropolitan Railway cost £186,000 per mile, a figure which has been quoted by some sources as rising to £1,000,000 in various places.

New York's 'EL' and a similar one in Chicago delayed subway building in the USA and the permission granted to Beach to build further subways for use by steam trains was never acted upon. However, it was inevitable that the pattern of progress would eventually demand their construction and it was not long before the elevated railway was unable to cope with all the traffic demands. So the New York Subway, or Inter-borough Railway, was constructed to extend from end to end of the island of Manhattan. Built just below the surface of the main streets, there were four lines initially, the inside one being used by trains that stopped only at principal stations. Passengers would travel by these as far as possible, then change to a slower train on the outside tracks which stopped at all intermediate stations. The subway lines came up to the surface to connect with the elevated line to cross Brooklyn Bridge, but as huge populations began to grow up on both sides of the rivers, the ferry steamers and bridges did not provide adequate communication. Tunnels were built under the Hudson and East rivers to form underground railway links between the mainland and Manhattan and from there to Long Island. Like the London tubes, they were constructed using the shield system to drive the tunnels. In one place the tubes had to be driven through 1,800 feet of solid rock.

The noisy and disfiguring lines of the 'EL' railways were abandoned finally in 1955, although some sections of the route have been adapted as special car 'speedways'. The New York subways now carry three-quarters of all the city's population who use any form of public transport. The lines radiate from the busy commercial and business centres of Wall Street and Fifth Avenue–42nd Street and are underground only in the central part of the city. Beyond this central area they are elevated above ground level. Most other main cities in the United States have constructed underground railways, notably Chicago where only 20 per cent of the total traffic of the public system is attributable to the subways and San Francisco whose Bay Area Rapid Transit District system is part underground,

part at ground level and part elevated. Both Chicago and San Francisco have recently constructed in-city railway lines along the median strips of motorways.

One of the most famous subway systems in the world is the Paris Métro. The line from Porte Maillot to Porte de Vincennes opened in 1900 and became the first railway to cross the city. Previously Paris had been serviced by surface radial lines and belt lines that encircled the city. The Porte Maillot to Porte de Vincennes line was followed by feverish construction so that within the next few years, the city was crossed by nine traverse, three radial and two circle underground passenger lines. Branches were also laid from the radial lines along the line of the old fortifications. Joined one to another by still more branches, these branches form a belt line known as the Petite Ceinture, which although it lost some of its significance with the subsequent development of road transport, is still used for trains linking the Gare de Nord and Gare du Lyon. The Métro network is so comprehensive that no part of the city is more than a quarter of a mile from a station, and stations on average are about a third of a mile apart. The efficiency of the system pays off for it is one of the most intensively worked in the world (although the New York City Transit Authority is the busiest subway in the world) and it handles considerable freight as well as passenger traffic. In one year, 1945, the Métro carried 1,508 million people as against the 201 million who travelled on road transport.

The twentieth century has seen the construction of subway systems all over the world. Berlin, Philadelphia and Boston opened their first lines in the decade between 1900 and 1910; Madrid, Buenos Aires and Oslo between 1920 and 1930, and Athens, Sydney, Stockholm and Moscow between 1930 and 1940. The Moscow Underground is particularly noted for the magnificence of its stations and its four lines – one of them circular – carry some very heavy traffic. London Transport operates the most extensive system in the world with over 250 route miles, although New York's network has considerably more stations as they are much closer together.

On the whole, underground railways are run with a higher degree of automation than those above ground. Such things as ticket control have long

Above: The Regional Express line in Paris runs east–west across the city. For convenience of passengers it is linked to metro stations along its journey, but its aim is to provide rapid travel from one side of Paris to the other.

Left: A shiny train pulls into an immaculately clean station on the Berlin Underground Railway. The station is Rebberge and the train is bound for Templehof.

been handled automatically in many parts of the world, and even more sophisticated methods in which computers are in charge of all fare charges and mechanisms are planned. The other area for automation is in the actual driving of the train, which can be done either by using trackside controls that feed information to individual trains or by a central computer that issues starting, stopping and running-speed instructions to all trains in the system, in accordance with pre-set programmes. This system has been used in San Francisco, but it is generally considered that passengers who travel on the underground railway would still rather their trains were manned by men than computers.

Above: One of the new Piccadilly line trains.

Right: An ex-Metropolitan railway locomotive – a 2-6-4 tank engine No 114 pictured at Amersham where it was still in use until 1938.

Left: One of the stations on the Moscow metro, which are well known for their extravagant architecture and cleanliness. The Moscow metro is the most highly developed of any underground rail system in Russia and is said to be similar, albeit with significant improvements, to that of London. It consists of a circle line which skirts round the city centre, three radial lines which cross the city and connect close to the Kremlin and three lines which lead out of the city from termini that are close to the city centre. In the 1960s, three million passengers travelled each day on the Moscow metro with trains running on the busiest lines at one-and-a-half minute intervals.

LONDON TRANSPORT

DECORATION AND BUILDING IN THE GREAT RAILWAY ERA

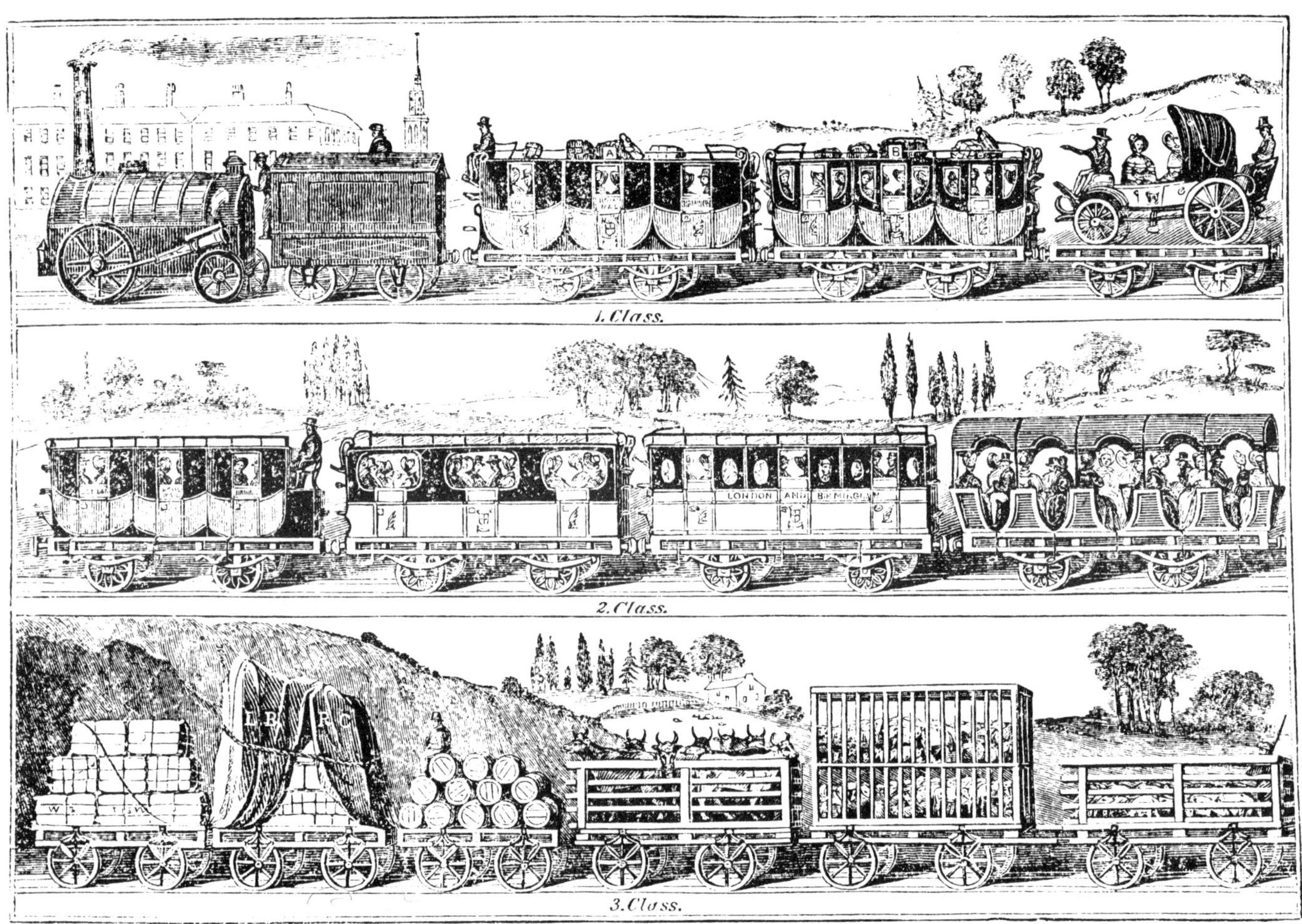

Railways evolved in the first instance not to carry passengers but goods, most usually coal or stone. Very occasionally the colliery or quarry companies who mainly owned the railways would put on a special 'excursion' train for the employees and their families. The passenger accommodation consisted of a 'cleaned-up' coal wagon, fitted with a few wooden boards for seats. Such conditions, combined with the smoke and grime issuing from the engine, must have had disastrous results on the finery that the passengers doubtless donned for such an auspicious occasion!

It was George Stephenson, in his fight to get steam locomotive power accepted as a form of rail travel, who conceived the idea of producing a 'carriage' for passengers, as opposed to the crude conversion of goods wagons. His aim at the time was to show that railway traction had other uses than just the conveyance of coal or goods. When the Stockton to Darlington Railway was opened in 1820 his engine, 'Locomotion', in addition to coal-laden wagons and open trucks fitted with seats,

Below left: Examples of first, second and third class travel on the London and Birmingham Railway in the late 1830s. The first class carriages were almost exact replicas of road stage coaches; second class coaches often had open sides and third class travel was clearly best avoided!

Right: George Pullman was the first person to bring any measure of comfort to rail travel. This is an example of a typical interior of a Pullman saloon carriage of the 1870s.

Below: A third class coach introduced on the Great Western Railways in the mid 1850s. These at least offered some protection from the weather.

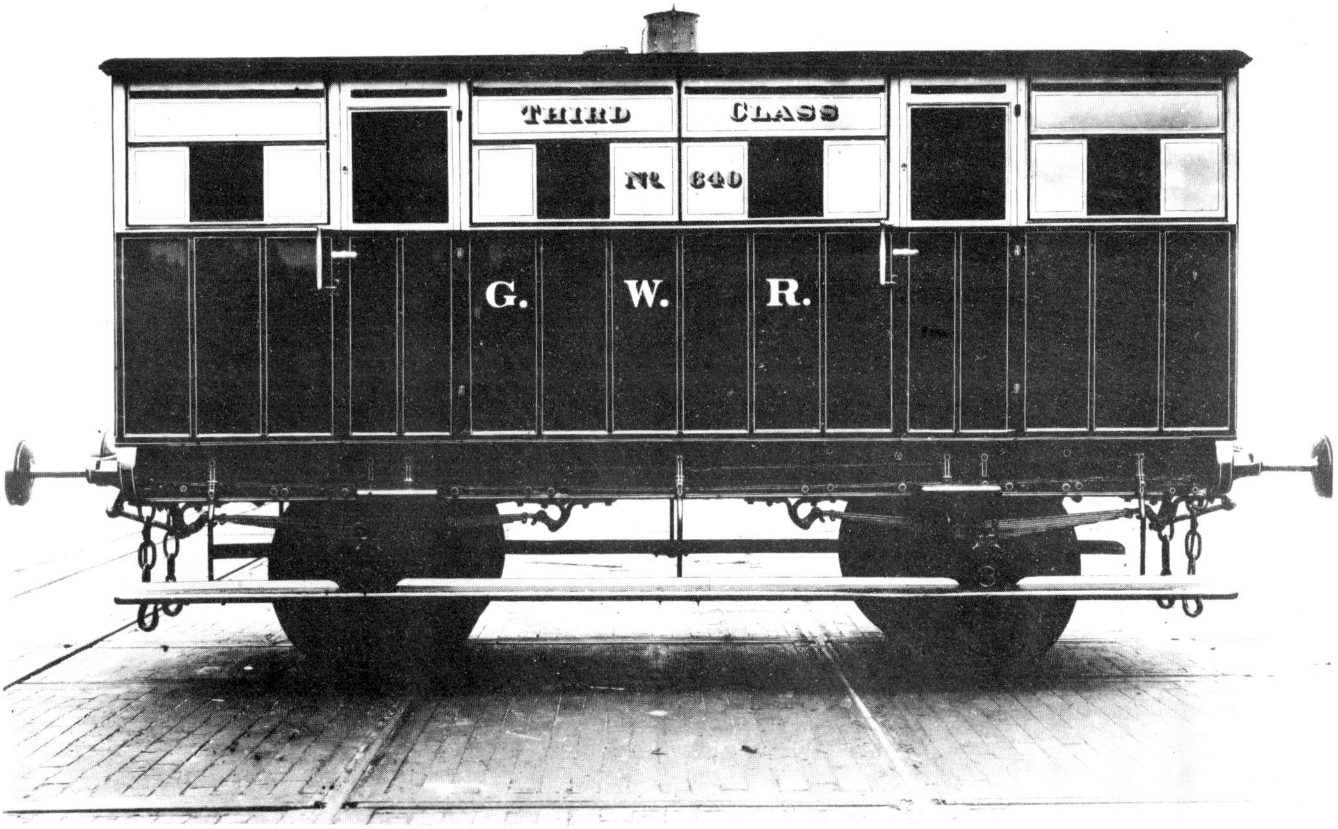

pulled a passenger carriage specially made for the occasion. Early reports of it indicate that it was little more than a 'shed on wheels', the entrance and exit positioned at one end. Undoubtedly the Duke of Wellington's carriage, pulled by Stephenson's locomotive 'Northumbrian' on the opening run of the Liverpool and Manchester Railway in 1830, was an example of more elaborate passenger accommodation. Contemporary descriptions of this say it was a 'gold and crimson vehicle resembling a Greek chariot'!

The Liverpool and Manchester Railway was probably the first in the world to develop passenger carriages, although in its early days many people actually travelled in their own carriages run onto flat trucks. The carriages built by or for the railways themselves, particularly for the first-class passengers, look just like the road-stage coaches of the day. Hilaire Belloc described them as 'a series of stage-coach insides tacked onto one another'. The first-class carriage seating eighteen passengers was in effect three stage-coach bodies mounted on a

railway-wagon frame. A slightly less elaborate but similar design was adopted for second-class passengers, although their carriages were frequently built with open sides. In both cases, luggage was put on the top of the coach, and in the absence of a guard's van, the guard sat at the end of one of the carriages. First-class passengers had cushioned seats to make travelling a little more comfortable, but no carriages had heating or lighting in the early days, although smoky and far from effective oil lamps were soon installed in the compartments. Most principal trains had first- and second-class passenger carriages only, but a few included third-class coaches as well. Third-class travel usually meant riding in open vehicles. Later, when covered wagons were introduced for third-class passengers, they only had tiny windows or ventilators placed so high that the passengers seated on the plain wooden, backless seats could not see out of them anyway. If lighting was provided at all, it was in the form of one dingy oil lamp hung over the division between two compartments to 'illuminate' both of them. Inducement to travel third class was further reduced by the fact that the trains would stop at every station along the route, so each journey became a very protracted, as well as uncomfortable, affair.

Advances in carriage design did not really come about until people began to experiment with formats other than the stage-coach prototype. The first designers to do so produced boxlike carriages of four or five compartments mounted on four wheels and constructed, it was said, 'like monks' stalls in old churches, so that none who sat in them should be comfortable enough to sleep'. In spite of their shortcomings, however, they represented a step in the general direction of all future carriage design.

A problem the early designers had to overcome was that of length. As the wheels were rigid, two or at the very most three pairs were all that could be

Above: The Midland Railway was the first to use the Pullman type of car, brought into this country from the USA in 1870. In spite of the doom-laden forebodings of some other railways who claimed that in the event of an accident the greater length of the Pullman car would crush the smaller cars to pieces, the Midland Pullman started a universal trend towards more comfortable and luxurious rail travel conditions in Great Britain.

Right above: The exterior of the Midland Railway's Pullman car.

Right: A long Buffet car belonging to the Chicago, Milwaukee and St Paul railroad. As with superior passenger comfort, USA was ahead of Great Britain in making refreshments available to rail travellers during their journey.

used if carriages were not to become too long to go safely round the curves of the track. This did not allow them to be built sufficiently long for general requirements. The 'bogie truck' was to provide the answer, enabling the wheels to pivot slightly, so curves could be safely negotiated. This enabled carriage builders greatly to increase the length of carriages.

It was generally conceded that the best ordinary carriages of any railway company in the early days were those of the Great Western. Its six-wheeled, four-compartment carriages have been described as resembling 'refined cigar boxes' as they were half as wide again as they were high – possible of course because of the wide gauge of the Great Western Railway. They were infinitely superior to any carriages found on narrower-gauge railways. Notable carriages of the Great Western at this time were the first eight-wheeled carriages ever, with four second-class and three first-class compartments. They were nicknamed 'Long Charleys'.

The first important attempts at sleeping cars

Left above: The dining saloon and 'library' on a turn of the century train of the Great Trans-Siberian Railway.

Left below: An early Pullman sleeper, showing one of the berths made up. Note also the oil lamps, spittoons and wicker seats.

Above: A photograph of a European restaurant-salon taken about 1908.

Right above: A first class compartment of a London, Brighton & South Coast Railway train in about 1908.

appeared in Britain in the early 1870s. They had in fact been preceded by 'bed-carriages' which had been introduced on the London and Birmingham Railway in the 1830s, with coaches that comprised a half compartment, two full compartments and a small additional boot. The backs of the seats against the boot were hinged so that they could be lifted and poles with webbing between them were placed across the space between the seats, with a large cushion placed on top. Passengers could then lie across the seats with their feet in the boot. The later sleeping carriages, which made their first appearance on the North British Railways, were rather less crude in design, and were similar to those finding popularity in France. The carriage was divided into two compartments separated by a short central corridor where there was a lavatory. The carriages themselves had two or three armchair seats with very high backs and by pulling a handle at the top of the back panel, the whole back came forwards to make a long bed with the daytime seat folded underneath. These sleeping compartments were quite elaborately furnished with ebony panels and plush velvet upholstery, but no bedding was provided. Passengers could bring their own with them if they felt they needed it!

These sleeping carriages were followed by others on the London and North Western Railway that comprised 'family salons'. They were long carriages divided into three compartments, two of which contained upper as well as lower berths.

The real advances in carriage design, however, came from an American who had been active in the United States for about fifteen years before his carriages reached Britain. His name – Pullman – is still synonomous with luxury rail travel. George Pullman began his successful venture into carriage design after experience of the discomforts of rail travel by night on the American trains. He persuaded the Chicago and Alton Railroad to let him convert three of their standard coaches to sleepers by giving the seats backs that could be let down, and by installing upper berths. It was some six years later, in 1865, that he patented his first design, which was the prototype of all future Pullman cars. By this time his first coach, named 'Pioneer', was in service; its first journey was with Abraham Lincoln's funeral train that ran from Chicago to Springfield.

The early Pullman coaches contained upper berths, 6 foot long that hinged up against the sides of the roof for daytime travel. Beneath each one was a pair of cross seats that faced each other. At

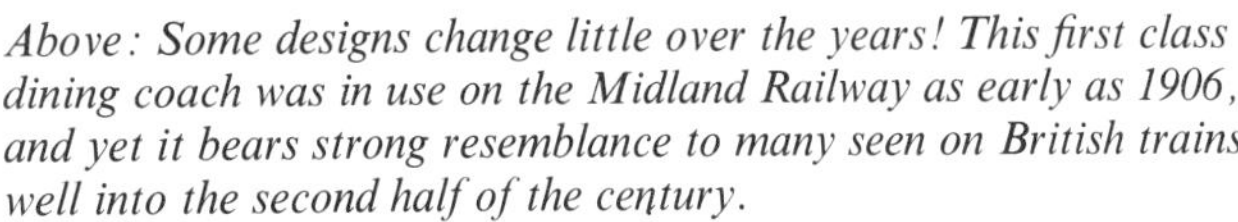

Above: Some designs change little over the years! This first class dining coach was in use on the Midland Railway as early as 1906, and yet it bears strong resemblance to many seen on British trains well into the second half of the century.

Above right and right: First and third class compartments of the Midland Railways in 1920. Apart from individual arm rests, the first class passengers do not appear to have had considerably more comfort!

Far right: Kitchen on a train. Meal preparation takes place on a Norfolk Coast express early in the twentieth century.

night these were pulled forward over the foot space to meet each other and the backs lowered to a horizontal position to fill the gap at either end. The bed was completed with a mattress and bedclothes. The sleeping sections were divided by panels that were raised from slots in the seat frames and the upper berths were similarly screened from one another by more panels that were slid into grooves. Heavy curtains buttoned together to partition the sleepers from the aisles, thus giving the passengers every privacy. We are told they were expected to remove their boots, and then take off their clothes while lying on their beds – a practice that apparently continued for almost a century thereafter!

The American public, quite understandably, were delighted with the new travelling luxury and were more than prepared to pay the $2 supplement charged. The cars' introduction was probably very timely, for it coincided with a period of general expansion of American railroads, and with the beginning of many long-distance journeys such as those of the Union Pacific and Central Pacific Railroads.

The first company to bring Pullman cars to the railways of Britain was the Midland Railway. It imported the standard parts that comprised the carriage and assembled them at Derby in the mid-1870s. The first carriage to appear on the company's regular service was appropriately named 'Midland' and apart from being built to slightly smaller clearances, it looked wholly American. The car's construction was very different from any of its British contemporaries, mainly because it was built in one piece as opposed to being two distinct units, which occurred when the carriage was mounted on a separate underframe.

The introduction of Pullman cars on the Midland Railway coincided with the abolition of second-class travel. All its trains now offered first- and third-class fares, but it was only the privileged first-class passengers who could enjoy Pullman luxury. Other companies, notably the Great Northern Railway, also began to incorporate the Pullman sleeping carriages into their regular service.

American trains by this time were not only providing their passengers with bed-time luxury. Sumptuous drawing-room or 'parlour' cars, smoking lounges, dining cars and various other special-purpose cars had been devised, although many of

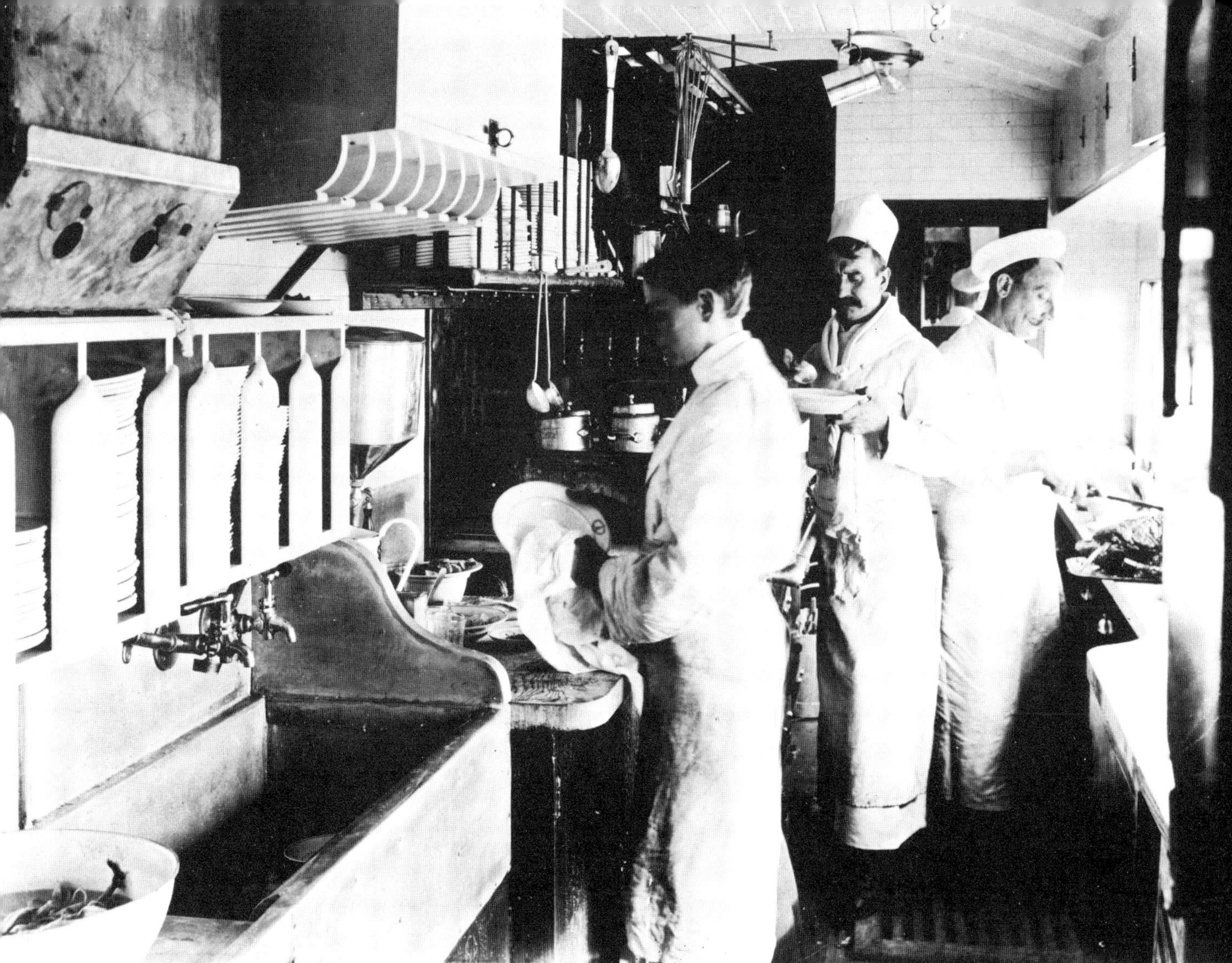

the railways in Britain had also found it necessary to provide some sort of smoking compartment. The Eastern Counties Railway had introduced open carriages entitled 'smoking saloons' or 'divans', with a central table with seats round the side, in the 1840s. The Midland Railway was the first again to bring parlour cars into use with a coach named 'Victoria'. It had a long general saloon with seventeen pivoted easy chairs; the original number of nineteen was reduced so as to give the passengers more leg room! At one end there were two reserved compartments which each had two armchairs and a sofa.

For a long time after rail travel had become the accepted, and was probably the most popular, form of general transport, refreshment and eating arrangements on long journeys were provided only at stations. The Great Western, for example, had an agreement with the refreshment room at Swindon, whereby they had to stop for at least ten minutes at this station. Brunel's derogatory description of the coffee served at the refreshment rooms at Swindon might interest rail travellers today, for things seemed to have changed little. He described it as tasting of 'roasted corn'! Trains en route to Scotland had twenty-minute 'refreshment' stops at Preston on the West Coast run, at York on the East Coast and Normanton on the Midland route. It was claimed that the refreshment rooms at these places served the soups so hot, that only the regular and experienced passengers were finished in time to be able to eat all the rest of the meal as well! Towards the end of the 1870s, however, the Great Northern Railway rebuilt one of their standard parlour cars as a drinking car. It had a kitchen at one end with a coke-burning range separated from the main dining compartment which had tables set between the standard parlour chairs. A smoking room was divided off at the other end and had a little stretch of side corridor. Lunches and dinners were cooked by a private caterer on the London to Leeds business expresses and were said to be 'memorably good'! The first dining car in the world was provided by the Great Western Railway of Canada in 1867.

With the advent of Pullman coaches in Britain came also the first 'corridor' trains. Hitherto, trains always comprised a series of coaches which were entirely separate and had no means of communication with one another while travelling. As may be imagined this suited the never very sociable British

Above: A special coach built in 1842 for the London & Birmingham Railway for use by the Dowager Queen Adelaide.

Right: Not just a carriage but a complete engine built in 1859 for the Viceroy of Egypt.

Opposite above: A carriage built in 1894 for King Oscar, of Sweden and Norway. He apparently used it mainly as a reception room.

Opposite below: 'The Duke of Connaught' – a royal train as can be seen from the coat of arms on the front – at Pretoria Station, South Africa in 1910.

travelling public which not only values it privacy, but is even now still loath to talk to fellow travellers! Corridor trains retained separate compartments but enabled passengers to walk from one end of the train to another, thus giving access to other compartments as well as the dining car, and so on. They were by no means greeted with acclaim at first and it took some time for their obvious advantages to be recognized.

Other countries of the world have taken rail travel to even greater degrees of luxury than was found in Britain, so that in some places trains virtually resemble travelling hotels. Long-distance trains such as the trans-Siberian express in Russia, or those that travel across the United States and from coast to coast in Canada, have sometimes incorporated hairdressing salons, shops, libraries, shower and bathroom compartments, boards which display up-to-the-minute information received at intermediate stations of stock-exchange news and racing results, typewriting compartments complete with secretaries, and nursery cars equipped with toys and nurses to keep children amused on long journeys – all of course in addition to the sleeping, dining and smoking cars. Scenic routes often had 'observation' cars where passengers could go to 'enjoy the scenery' or obtain some fresh air. Trans-Australian trains even provided pianos for the amusement of their travellers. There have also been trains in out-of-the-way areas that carried church cars containing a chapel and accommodation for missionaries. Elaborate provision was often made for the transport of livestock and special agricultural trains used to tour rural areas with exhibitions of modern farming equipment and lecture facilities on new farming methods. Postal vans did not merely convey mail; they were equipped with sorting racks so that letters and parcels could be handled 'on the

No 671

move' and, in some instances, they were delivered en route.

Sumptuous carriages have been built in the past for royalty and other state dignitaries to make travelling conditions for these eminent personages as comfortable and pleasant as possible. Napoleon III had a royal train of richly appointed carriages with 'a wine cellar and a conservatory of flowers'. This train was later bought by the then Tsar of Russia who had it enlarged and made into a 'palace on wheels'. The Austrian Emperor Francis Joseph was presented with a special train of eight magnificently decorated cars by the administration of the Austrian railways. Rulers of other lands, the princes of India in particular, have had luxurious private trains on which comparative fortunes were spent to provide every imaginable luxury. Some ended their lives considerably more ignominiously than they had begun. The elaborate, eight-wheeled, compound-type bogie carriage built by the American firm Mason and Company for the Viceroy Said Pasha of Egypt in the late 1850s later became a departmental mobile laboratory in Upper Egypt, and a private, iron-panelled car built in the early 1860s for President Lincoln was burned in a grass fire in the American Middle West some years later.

Provision for royal travel on the railways of Britain has always been provided by the individual rail companies, rather than by the royal household owning its own private train. Members of the royal family would often travel as private passengers in a specially appointed saloon attached to an ordinary train, and would pay a normal fare to do so like any member of the public. For special journeys, however, the royal party, which might comprise any number of ministers of state and other VIPs, as well as a large party of ladies-in-waiting and gentlemen attendants, would take over a complete train which then, not unnaturally, was designated as the royal train. Several railway companies kept special locomotives to pull the royal trains. The Great Eastern's engines, for example, were at one stage painted dark blue and lined with red. A few also had white lines painted on the body work, and they were considered to be suitably patriotically adorned to pull the royal train! The names of some of the Great Western engines – 'The Queen', 'Empress of India' and 'Royal Sovereign' – tend to belie their original use, although these locomotives were also used for hauling regular trains at other times. For the royal journey from Euston to Carlisle in 1897 – the year of Queen Victoria's Diamond Jubilee – a bright-red engine called 'Greater Britain' took the train to Crewe, where the 'Queen Empress', a cream-white engine, took over to complete the journey to Carlisle. As may be imagined such engines were kept in pristine condition, painted and repainted whenever

Above top: The exterior of the Royal Train built in 1903 for Edward and Alexandra showing the individual carriages. It provided sleeping and dining accommodation and was used mainly for regular journeys between Euston and Balmoral.

Above left: The sumptuous interior of a coach provided for Queen Victoria in 1869.

Above right: The bathroom leading through to the bedroom in the King's salon.

Left: The bedroom of the Queen's salon.

necessary, and besides carrying the royal coat of arms, they were often decorated with flags and flowers for their journeys. Stations through which the royal trains had to pass would be similarly decorated and would often have a military band playing.

Many special carriages have been built over the years for royalty. One of the first was the royal saloon for the Dowager Queen Adelaide. Built in 1842 by the London and Birmingham Railway, the body of the coach was made by the most distinguished coachbuilder in the country, the firm of Hooper, which was later to make bodies for the Rolls Royce company. It was painted a 'deep claret' colour with the royal coat of arms displayed on the side. It had a sleeping compartment as well as a daytime saloon and was lit by two oil-pot lamps.

Queen Victoria was said infinitely to prefer travelling on royal trains to the royal yacht, and during her long reign many carriages were built for her private use. An early one, in existence by 1844, for use between London and Gosport, was in fact still in service a hundred years later, although not by then as a royal carriage. It was a modification of the usual three-compartment type and the space normally occupied by the middle and end one formed the main saloon. It had four armchairs and a fixed sofa. The single compartment at the other end, which had an internal connecting door as well as its own outside doors, was a sort of ante-room with a sofa and two corner seats. Both ends of the carriage had large observation windows. The interior decorations were very luxurious with silk and lace upholstery on the seats, silk and velvet padding on the ceiling with silver decorations, silk curtains and deep expensive carpets.

At the beginning of the 1850s, Richard Mansell designed the most elaborate royal saloon so far run on any of Britain's standard-gauge railways for the South Eastern. It had three compartments, the middle and largest one forming the state saloon, with an adjoining door leading to a vestibule at one end. The attendants' compartment was at the other end. According to Mansell's description, it sounds as if the interior décor was unrivalled in its opulence, comprising rich furnishings of silk, satin and velvet, specially carved chairs, fine inlaid tables and deep-pile carpets.

The places to which Queen Victoria most frequently journeyed dictated that she would travel mostly on the London and North Western (with the Caledonian and the Great North of Scotland), the Great Western, the South Western and the South Eastern Railways. All these companies therefore had either complete royal trains or at least one royal saloon, such as the luxurious one already described belonging to the South Eastern. The Great Western Railway built a special Diamond

Jubilee train in 1897 but on royal command direct from the palace, it had to incorporate the Queen's Carriage they had built in 1873. This had found great favour with Queen Victoria, and although it was lengthened to enlarge the ante-rooms at either end, the central state compartment was unaltered in accordance with her wishes. Indeed from descriptions of the interior, which once more was a supreme example of Victorian opulence, there seemed little need so to do! The Great Western's royal train consisted of seven coaches in all and was the only royal 'corridor' train during Victoria's reign.

The Victorian opulence appeared to find considerably less favour with the great lady's successor, Edward VII, whose instructions, on being consulted about the building of a new royal train shortly after he became king, were to 'Make it as much like a yacht as possible!' Consequently the new saloons of the royal train built for Edward and Alexandra in 1903 displayed interiors of gleaming white enamel. The train had day and night saloons for both monarchs as well as various offices, ante-rooms, dining saloons and smoking compartments. It is perhaps the most famous of all British royal trains and was used after Edward VII by George V and later George VI for most of the all-night royal journeys. In fact it was probably among the last of the highly luxuriously appointed royal trains, for in more recent times royalty, and eminent people the world over, travel in carriages that by no means echo the palatial splendour known to their ancestors.

No less fascinating than the development of railways and the locomotives and trains that ran on them has been the building of railway stations, particularly the great termini. Stations in the early days, before there was a need for complicated signalling equipment and communication with other places, were little more than wayside stops but the public's demand for greater speed and comfort on trains as well as more frequent and improved services became inevitably coupled with similar requirements for improved conditions and facilities at stations.

Accordingly the great stations of the world began to be built and many of the wooden shacks of former times were replaced by amazing and massive structures. In Britain, Brunel for example designed a superb timber roof constructed of arches for the terminus at Bristol's Temple Meads. Bath Spa station was approached by a viaduct of 73 arches and the entrance to Shrewsbury station looked like an Elizabethan mansion. York provided a good example of a station that was developed so as to adapt to the changes in rail travel when it became necessary to provide through accommodation from

Above: The Midland Railway Hotel attached to Euston was surprisingly not the most impressive or well-known construction attached to the station. This title went instead to the Doric Arch which in fact marked the entrance to the station yard.

Right: An etching showing Queen Victoria at York Railway Station in 1849. Later, York Station was almost completely rebuilt in order to provide through north-south communication.

Far right: The outside of the terminus at Bristol Temple Mead Station in 1850. Behind the façade the most imposing feature was the timber roof which was constructed of several arches.

north to south for the trains of the York and North Midland and the Great Northern. A new station was built comprising huge concentrically curved 'train sheds', which had five carefully scaled and centred ribs.

It was in London, however, that the really impressive British stations were built, for they were necessarily the largest. Euston became the most famous, and most frequently illustrated, with its amazing Doric archway which was erected at a cost of £35,000 and might be the entrance to a museum rather than merely the station yard! St Pancras Station, built as the London terminus for the Midland Railway in 1868, was designed by the architect Sir Gilbert Scott. Its great Gothic-shaped roof, built of glass and spanning 240 feet rises in an unbroken line from platform level to meet at the top central ridge 100 feet above rail level. In this station the platforms were in effect positioned in the roof for there were two storeys beneath them used for entirely separate purposes. Waterloo, originally built for the London and South Western Railway (later incorporated in the Southern Railway) replaced a former wilderness of wooden shacks and sheds. Nowadays it is the station with the largest number of platforms in Britain.

It became common practice in the second half of the nineteenth century to build huge hotels at the London termini and many of these remain today, even if their former Victorian glory is somewhat tarnished. The first one was built at Paddington in 1854 and formed the main entrance to the station, as similar buildings also did subsequently at St Pancras and Marylebone.

It has been claimed that the great stations in Britain deal with such a huge amount of traffic in such small areas that there is not room for the splendid grounds and gardens found in stations complexes elsewhere in the world. A beautiful example of a garden station was that of 'Southern' station in Vienna, which was built to appear more garden than station! Stations in the United States often fall into a similar category and if they do not all have carefully tended gardens, they often do have attractive concourses in front of the station proper. A splendid example of this was the Washington Union Station which was built with thirty-two platforms, each one entered by a separate gate positioned in a long ornamental ironwork barrier. It also had a 'state entrance' leading to a suite of waiting and reception rooms designed for the use of the President. The magnificent grounds in front of the station echoed the importance of Washington as the seat of government and it was made so large because of the enormous crowds to be dealt with at Presidential elections.

Two of the most magnificent stations of all are the Pennsylvania Terminal and New York's Grand

Central Terminal; the latter is in fact the largest railway station in the world. They have both been described as 'veritable palaces' and have arcades of shops and restaurants as well as hotel and office complexes. The many fine railway stations in America owe something of their great scale to the fact that they were often built by the joint forces of several railway companies so as to provide one huge station instead of several smaller ones. When the Chicago and North Western Railway built its huge terminus it was claimed as one of the most expensive and luxurious the world had ever seen. It comprised private suites of apartments for ladies and children – including bedrooms, tea-rooms, bath and dressing rooms, as well as restaurants, nurseries, first-aid rooms, gentlemen's dressing, lounge and smoking rooms, chemists, hairdressers and manicuring and boot-cleaning services!

Elsewhere in the world stations of similar magnificence have been built. In India many of the great stations and railway-office buildings resemble palaces, particularly those built for the original Great India Peninsular Railway in Bombay and the East India Railway in Calcutta.

Canada has several big stations, of which a famous one was the Windsor Street Station in Montreal, built to resemble a French château. Continental Europe too has many fine examples of railway station architecture (Vienna has already been mentioned) and they are often noticeably less

utilitarian in appearance than those of Britain. The Anhalter station in Berlin, built in the latter part of the nineteenth century to the designs of the architect Franz Schwechter, is considered by some people to have the finest frontage of any terminal station in Europe. The entrance porch projects forward, while behind it soars the vast curvilinear mass of the 'train shed'.

Another station claimed as one of the most impressive in Europe when it was opened in 1931 was the Central Station in Milan. The approach to the main building was through three great portals from the Piazza Andrea Doria along an imposing covered carriage drive. The booking hall which opened off the drive presented a striking aspect of stained-glass windows and multi-coloured marble and stone.

Stations built today rarely display such design extravagance. Like almost everything to do with railways, romance has given way to utilitarianism. Thus carefully tended, much-loved, highly polished locomotives with their luxurious carriages have been replaced by characterless-looking engine units that pull stereotype trains. Modern stations tend to be huge glass and concrete structures built for efficiency rather than aesthetic appeal. To those who know and loved the era of steam-locomotive travel and all its peripheral characteristics, modern rail travel must probably seem a high price to pay for progress.

Above left: In marked contrast to the splendour of the large railway termini, this simple country station mark's a familiar sight to all rail travellers in Great Britain.

Above top: The 'coffee room' at the St Pancras Hotel, in 1876.

Middle and bottom: Two shots of St Pancras Station, built in 1868 as the London terminus to the Midland Railway. Before this the railway's access into London was over 32 miles of the Great Western tracks.

INDEX

Numbers in italics refer to the page numbers on which illustrations appear

Acknowledgments

The publishers would like to thank the following individuals and organizations for their kind permission to reproduce the photographs in this book:

American History Picture Library: 35 above, 36; W. J. V. Anderson: Back Jacket; Australian News and Information Bureau: 140 below; B. T. Batsford Ltd.: 15 below, 179 right, 181, 188-189 above; British Rail: 64-65 below, 148-149, 158 above, 159 above, 176, 180, 182 above, 184-185, 189, (Haresnape) 19, 59, 67 above, 68, (Ian Yearsley) 147 below; British Tourist Authority: 10 above, 21; Canadian Pacific Railway: 152 below, 154-155, 157; C. C. Q. (R. M. Quinn): 136 above right, 173 below; C. R. L. Coles: 72 above; Colourviews: 26-27, 58 below, 62 below, 65 centre, 66; John R. Day: 168 above and centre, 169 above, 170-171; English Electric Co.: 164-165; Mary Evans Picture Library: 174, 187 below; C. J. Gammell: 23, 24, 56, 63, 72 below, 106, 107; Victor Hand: 41, 44 below, 45, 47, 50 below, 51, 52, 55, 73-88, 90-91 below, 92-99, 100-101 above, 102-104, 108-127, 128 above, 129-134, 136 above left, 136 centre and below, 138-139, 144-145, 146-147, 146 below, 153, 156; Haresnape: 17, 20 above, 20 below, 70-71, (Ian Allen) 57, 58 above, (Hebron) 65 above, 67 below; Heyer/Henn Fotograph/P. Winding: 139 above; Indian Railways (Ian Yearsley): 101 below; I.P.S. Photo: 137; Japan Information Centre: 159 below; Jim Jarvis: 44 above, 46, 48-49, 177 below, Front Jacket; R. H. Kindig: 53, 54; London Transport Executive: 163, 165, 167 below, 168 below, 169 below, 172-173 above; Howard Loxton: 32-33 below, 34-35 below, 179 left, 183 above; Mansell Collection: 175 above, 186-187; Milwaukee Road: 38, 39; New South Wales (Government Office): 89, 90-91 above; Novosti Press Agency: 172 left, 178 above; Ed Nowak: 50 above; Picturepoint: 105; Radio Times Hulton Picture Library: 11, 20 centre, 25, 32-33 above, 34 above, 37, 62 above, 69, 138 left, 141-143, 160-162; Science Museum: 6-7, 8, 9, 22, 42-43, 175 below, 177 above, 182 below, (Haresnape) 10 below, 12-13, 15 above; SNCF (Peter Winding): 150 below, 158 below; South African Railways: 183 below; Union Pacific Railroad Co.: 178 below; U.S. Information Service: 28, 30-31, 166-167 above; Peter F. Winding: 138 above right, 140 above, 150 above, 151, 152.